BE THE CHANGE

YOU WANT TO SEE

IN THE WORLD

365 Things You Can Do for Yourself and Your Planet

Julie Fisher-McGarry

Foreword by **John Robbins**, author of *The Food Revolution*

D0359417

Conari Press

To my mum, Elsie Mabel Fisher.
You are forever in my heart.

First published in 2006 by Conari Press,
an imprint of Red Wheel/Weiser, LLC
With offices at:
500 Third Street, Suite 230
San Francisco, CA 94107
www.redwheelweiser.com

ISBN-10: 1-57324-297-7
ISBN-13: 978-1-57324-297-4

Library of Congress Cataloging-in-Publication Data

Fisher-McGarry, Julie. Be the Change You Want to See in the World: 365 Things You
Can Do for Yourself and Your Planet/Julie Fisher-McGarry.

    p. cm.
Includes bibliographical references and index.
ISBN 1-57324-297-7 (alk. paper)
1. Organic living. 2. Human ecology—Philosophy. I. Title.
GF77.F57 2005
158.1—dc22

                                2005017694

Cover and book design by Maija Tollefson
Typeset in Joanna

Printed in the United States

TS

10 9 8 7 6 5 4 3 2 1

Printed on 100% post-consumer recycled, chlorine-free paper

CONTENTS

FOREWORD

Most of us want to live healthier lives. Most of us want to be part of the solution and not part of the problem. But it's not so easy in a world that seems out of touch with some important and basic human realities.

That's why Julie Fisher-McGarry's book is so important, so useful, and so needed. It's simple, easy to use, and fun. Each day's entry is clear and uplifting, and will help align your life with your values and your love.

That's no small accomplishment in a world so out of balance.

I'm grateful for Julie's book, and proud of her for writing it. And you get to be the beneficiary of her efforts. Enjoy this book and put it to use. Your body and soul will thank you every day of this year, and every year yet to come.

—John Robbins, author of *Diet for a New America*,
The Food Revolution, and *Healthy At 100*

ACKNOWLEDGMENTS

Hugs to Vern for your encouragement in writing this book. Always know that I love you and will forever cherish the wonderful times we had together. It is said that we only truly love a very few people in our lives; you are one of mine.

And to Smiler and Camelot: precious forever companions.

Thank you to my publisher, Jan Johnson, and my editors, Jill Rogers and Caroline Pincus, for your vision of *Be the Change You Want to See in the World*.

Thank you to everyone else in this book, named and unnamed, for your inspiration and perseverance in making this a better world for all.

Health and Happiness.

INTRODUCTION

Be the Change You Want to See in the World is a daybook for helping us keep our hearts and minds open to all that lives. I wrote it originally for myself, in journal form, as a way of keeping an account of all the significant things I learn about as I go through life, information too detailed and important to allow to fade away. As my stack of journals grew, the writing began to take on a distinct shape. No longer just random notes, my jottings started coming together into a coherent collection of nonreligious yet deeply soulful suggestions for demonstrating concern, kindness, and consideration for nature and its animals in our daily deeds.

For every day of the year I offer personal anecdotes, inspiring quotations, verses of poetry, simple vegetarian recipes, women's health tips, environmental facts, and green thoughts—each just a small thing that you can do to have a positive impact on the world. Alone your actions are significant; together with others their impact can be enormous. You don't need any special tools or skills to begin building a better world, but I do encourage you to keep a journal as you read along so you can record your own green and compassionate ideas and plans and daily acts. When we see our words in print, either written by hand or on a computer screen, we can be clearer about what we are thinking, clearer about our intentions, clearer in our plans.

I wrote the book for women. Why? Because I believe that women are less afraid than men to show emotion and act upon it. We are more compassionate and sensitive by nature and embody a certain humility and humanity that predisposes us to want to make a difference. Of course I welcome men to read along as well; if there are entries that you are tempted to skip, think about sharing them with a woman friend. As I see it, we are a sum total of our thoughts and actions and we should put our energies toward the good.

By reading this book you will discover why individual acts really do matter: if each of us drops a large enough pebbles into a large enough pond to ensure large enough ripples, we can create a new world without hunger and pain, without war, injustice, or cruelty. I wrote this as a daybook to keep that pebble in the forefront of your mind; we *can* start small and make a big difference. As demonstrated so beautifully in the movie, *Pay it Forward*, a better life starts with one person helping another with a simple act of human kindness, which then inspires both that person and anyone who witnesses or hears about it to do likewise. If every reader brought just some of these thoughts and ideas to life, the world would be a much better place.

And it all starts now, with my heart and with yours. May your skies be forever blue and your thoughts forever green.

—Julie Fisher-McGarry

JANUARY

Awakening to the New Year

January 1 - New Year, New You

If you didn't receive a journal as a holiday gift, go out and buy yourself a gorgeous one. This will encourage you to write when you have the need—be it once a day, once a week, or just once in a while.

January 2 - Looking At Your Life

Take a look at your life. Is there any time in your day for you? If you need more breathing room, then cut back on your schedule and learn how to say a kind but firm, "No!" to colleagues, family, and friends. Find a way to stop doing an uncontrolled plunge down the rabbit hole each day and examine your life to find a solution. Don't try to change everything overnight. Just apply the brakes, gently. Take a breath, a deep one from the diaphragm, before you burn yourself out.

January 3 - Every New Year is a New Opportunity

January gets its name from the Roman god "Janus," the patron of endings and beginnings. He is shown as having two faces looking in opposite directions. So, like Janus, look back on the past year and be prepared for the next year.

January 4 - Living By Giving

If you feel you have no purpose in life, try feeling for someone else. Call on an elderly neighbor or single mom to say "Hi" and see if they need anything, even if just a smile. Volunteer at a nursing home, or an animal shelter to give yourself a sense of purpose, of being useful. You'll make new friends, you'll feel joyful and uplifted and will stop wallowing in self-pity.

January 5 - Cancer Proof Your Diet

The World Cancer Research Fund, the American Cancer Society, and the Royal Cancer Society in Britain—all organizations that study the issue—agree that as many cases of cancer are caused by diet as are caused by smoking, and all of them make the same top-two recommendations for preventing cancer: Eat more plant-based foods and eat fewer animal-based foods. In other words, "go vegan."

January 6 - First Steps to Better Health

If, however, you aren't doing too much but too little, try going for a brisk nature walk every morning. It'll wake you up and revitalize you. Use all your senses.

January 7 - Simple Soups

Hearty vegan soups are filling yet low in saturated fat, full of vitamins, protein, and fiber, and huge on comfort factor. Serve 'em up with a good quality crusty whole wheat bread. Then add a bowl of fresh salad. Make enough for two days, or pass half on to a neighbor or relative.

Here's one of my favorites:

Costa Rican Black Bean Soup

3	cups dried black beans, soaked 6-8 hours
10	cups water
1	teaspoon salt
2	sticks celery, chopped
1	onion, chopped
1	green bell pepper, diced
2	teaspoons dried oregano
1	teaspoon black pepper
2	tablespoons ground cumin
1	teaspoon crushed red pepper flakes
2	bay leaves
1	can (14 ounces) chopped tomatoes, undrained
1	can (8 ounces) tomato paste
2	tablespoons hot sauce

Drain, sort, and rinse beans and place in a large Dutch oven or stock-pot. Add water and salt, and bring to a boil. Boil for 10 minutes. Stir the celery, onion, bell pepper, oregano, black pepper, cumin, red pepper flakes, and bay leaves into the beans. Return to a boil; cover, reduce heat, and simmer 1½ hours or until beans are tender. Add tomatoes, tomato paste, and hot sauce, stirring well. Return to a boil; reduce heat and simmer, uncovered, for 30 minutes. Discard bay leaves. Check seasoning. Serves 8.

January 8 - The Cure for a Cold Winter's Day

My Minestrone

2	teaspoons olive oil
1	medium leek, chopped
2	carrots, diced
2	celery stalks, diced
1	clove garlic, crushed
6	cups water
1	cup cooked white beans
	coarsely ground black pepper, to taste
1	teaspoon dried oregano
2	tablespoons tomato paste
1	cup cabbage or broccoli, shredded
1	cup frozen peas
1	tablespoon fresh parsley, chopped
¼	cup whole wheat pasta, uncooked
½	cup vegan Parmesan cheese

Heat oil in a large saucepan over a medium-high heat. Add leek, carrots, celery, and garlic and sauté for 4 minutes or until softened. Add water, beans, pepper, oregano, and tomato paste. Stir and bring to a boil. Reduce heat to a simmer and cook for 30 minutes. Add more boiling water if it needs it. Add cabbage or broccoli, peas, parsley, pasta, and cheese. Stir and cook an additional 10 minutes. Serves 4-6.

January 9 - To Look Good is to Feel Good

January is a time of resolution and renewal. A chance to dream a new and wonderful dream and, like Janus, show a bright face. Showing that fresh face to the world comes from within, from an inner confidence, grace, health, happiness, and self-satisfaction. To look good you must feel good, or it will show in peevish lines around your lips, in tired eyes, or in tightened, dry-looking skin.

January 10 - Let's Check the Basics

- Are you getting enough sleep? Most people need eight hours; some need more.
- Are you eating properly, a diet low in fats and sugar, high in natural fiber, with six or more helpings of fresh fruit and vegetables each day?
- Are you drinking plenty of water?
- And not too much alcohol and caffeine?
- Are you breathing from the diaphragm to oxygenate your blood?
- Are you exercising, with at least one 20-30 minute brisk walk a day or the equivalent?
- Do you relax and take time for you? Yoga, meditation, daydream, a good book?
- Do you laugh and have fun every day?
- Do you have a rewarding job that is fulfilling and makes use of your talents?
- Do you have a happy, comfortable, and loving relationship with your partner, children, family, and friends?

- Are you living each day to its fullest, as though it were your last?

- Are you continually learning, growing, and experiencing life?

- Are you a wonderful work in progress?

- Do you need to work at any of these? I know I do! I'm sure we all do. Make observations of your own in your journal—write them down, read them, apply them to yourself and to your life.

January 11 - Skin Savers

If the blustery wind and indoor heating are leaving your skin gray, dry, and flaky, it's time to nourish it. Go get yourself a facial. If you don't have a regular salon, try visiting a few day spas and choose the one where you think you will feel most comfortable and cosseted. Ask about the treatments they offer and choose the one right for you, or ask for a recommendation.

You can also pamper yourself at home. Begin every day with gentle cleansing, toning, and a moisturizer for your skin type. Don't use any old soap and water. If you like that just-washed feeling on your skin, choose a gentle foaming cleanser that won't strip away your own natural oils. Once a week, use a gentle exfoliator to buff away those dead skin cells and an appropriate mask to deep cleanse.

Choose the right moisturizer for your skin condition; not just for whether it is dry, oily, or sensitive, but one that will help with firming, hydrating, tightening, or with wrinkles. If you spend time out of doors your moisturizer needs to have an SPF of at least 15, even in winter. I also recommend a good night cream that is heavier and will nourish your skin as you sleep.

Remember to use your moisturizer on your throat and neck twice a day also, as skin in this area is thin and can be the first to show signs of aging. And don't forget those eye creams. These are lighter and specific to dark shadows, puffiness, or laughter lines.

January 12 - Guilt-Free Beauty

Be sure to buy beauty products that have not been tested on animals. Check with *www.leapingbunny.org* for a list of companies that do not test finished products, ingredients, or formulations on animals, or you can phone 1-888-546-CCIC (Coalition for Consumer Information on Cosmetics) and they'll be happy to send you a pocket shopping guide of companies that manufacture with compassion. Or look for the leaping bunny logo on the product.

January 13 - Return to the Library; Return to Reading

My very first job straight out of school at sixteen was as a library assistant. Libraries are wonderful. Everything's free, the resources are nearly endless, and you'll save many a tree!

January 14 - Conscious Crafts

When I watch TV, I love to knit and crochet. Everyone I know already has a colorful throw I've given to them. So now what to do with the products of all that handwork? Web sites to the rescue!

www.woolworks.org has a section on knitting and crocheting for charity listed by state, with links for organizations that make donations to U.S. troops; to centers for battered women and children; to Native American Reservations in Arizona, New Mexico, and Utah; and to the homeless in cities all over the U.S.

www.newbornsinneed.org would love clothing and bedding items for newborn, sick, needy, and premature babies. You can donate blankets, hats, booties, and afghans. All they ask is that you use the softest yarn possible.

www.warmingfamilies.org is a 100% volunteer project that delivers donated blankets and other warm items to the homeless and displaced while strengthening families with their charity work. They supply to local shelters and nursing homes.

January 15 - We Are the Keepers of the Flame

The ultimate measure of a man is not where he stands in moments of comfort and convenience, but where he stands at times of challenge and controversy. The true neighbor will risk his position, his prestige, and even his life for the welfare of others.

Those words were spoken by Dr. Martin Luther King. Today is his birthday. Take a moment today and think about what Dr. King's life means to you. Go to *www.mlkday.org* for more information on this great man and for suggestions of projects you can get involved with to help your community. Honor Dr. King, not just this week, but every week, and do your best to help other people.

January 16 - Meatless Mondays - Good for You AND the Planet

Whether you are motivated by concerns for yourself and your family's health, ending unnecessary animal suffering, environmental protection, or conserving scarce global food resources for the hungry, Meatout Mondays are the answer. Visit www.farmusa.org to sign up for a great weekly e-mail with a recipe, an inspirational message, and an informative feature that will help you kick the meat habit.

January 17 - Relax

When your life gets chaotic, try to relax. Open the front door and go for a walk or a jog. Enjoy the fresh air—the rain if it's raining, the sun if it's shining. Breathe deeply and forget about that pile of work on your desk, just for twenty minutes. A walk will help lower your stress levels, your blood pressure, and your bad temper.

January 18 - Baked Goods That Are Good for You

January is a great time for baking. The nights are long, the days are short, and in most parts of the country it's cold and dreary.

Here is my favorite: hardly any cholesterol, with soymilk and tofu for calcium, dates for iron, and whole wheat flour for fiber and B vitamins. This recipe is easy, foolproof, and very flavorful—enjoy!

Spicy Ladies Date Cake

1½	cups pitted dates, chopped
1	cup vanilla soymilk
1	cup granulated sugar
6	ounces soft tofu, drained
3	tablespoons vegan margarine, softened
1½	cups all-purpose whole wheat flour
1½	teaspoons baking powder
½	teaspoon baking soda
1	tablespoon ground cinnamon
1	teaspoon grated nutmeg
½	teaspoon salt
2-3	tablespoons dark brown sugar

Preheat oven to 350°F. Coat an 8x8 inch baking dish with cooking spray. In a small saucepan, combine dates and soymilk. Bring to a boil. Remove from heat and let stand for about 5 minutes until the dates are soft. Then place in a food processor and whiz until smooth. Add the sugar, tofu, and margarine and process until smooth. In a large bowl, sift the flour, baking powder, baking soda, cinnamon, nutmeg, and salt. Stir in the date mixture, but don't over beat. Spread this batter into the baking dish. Scatter the brown sugar over the top, and bake for 30 minutes at 350°F, or until a toothpick inserted into the center comes out clean. Cool in the dish for 10 minutes, and then remove to cool on a wire rack, or leave in the pan. Cut into squares and eat until gone!

January 19 - Making Memories

Ginger Molasses Cake

1	cup whole wheat flour
1	cup unbleached white flour
1	tablespoon baking powder
2	teaspoons ground ginger
1	teaspoon ground cinnamon
½	teaspoon freshly grated nutmeg
¼	teaspoon salt
2/3	cup soymilk
1/3	cup safflower or any light oil
1/3	cup molasses
1/3	cup maple syrup
2	tablespoons fresh ginger, grated
2	teaspoons vanilla extract

Preheat oven to 350°F. Lightly oil an 8x8 inch baking pan. In a large bowl, whisk together the flours, baking powder, ground ginger, cinnamon, nutmeg, and salt. In a medium bowl, stir together the soymilk, oil, molasses, maple syrup, grated ginger, and vanilla extract until well combined. Tip these ingredients into the dry ones and stir together until blended. Plop into your prepared pan. Bake for 35-40 minutes or until an inserted toothpick comes out clean. Leave to cool in the pan before cutting into 9 pieces.

January 20 - Get a Move On

Try to make time for at least a twenty-minute cardio routine five days a week—somewhere, somehow. Set your alarm clock to go off a half-hour earlier and take a brisk walk or jog to wake

Be the Change You Want to See in the World

yourself up. If that's too much (or you know it wouldn't last), then break it into more manageable sessions: ten minutes first thing to rise and shine, a fifteen-minute walk at lunchtime, then hit the gym on the way home for some strength training.

January 21 - It Is Far Easier Being Green

Are you green? It's easy to get complacent or cynical. One person switching off a light for one hour may not make much difference, but one million people across the country will—and it's not just a case of your fuel bill, it's an important way to save and preserve our planet Earth.

January 22 - Energy Smarts

• Turn off lights when you leave a room.

• Turn off the heater or air conditioning when you don't need it, or use a heater or fan only for the room you are using.

• When doing the laundry, do a full load or use the economy setting.

January 23 - Save Water

• Fill the sink with water when shaving, brushing teeth, or rinsing, instead of letting the tap run.

• Share a shower or bath with a friend, or take quicker showers.

• Don't overwater the garden and lawn. Remember to turn off the sprinklers when it rains.

- Don't hose down driveways and walkways—use a broom.

- Minimize the use of waste disposals as they use a lot of water.

January 24 - Waste Not

- Use recycled items whenever you can.

- Separate paper, metals, and plastics for recycling.

- Reuse bags or take canvas bags to the market.

January 25 - Save the Air

- Support political actions for clean air.

- Avoid using harsh chemicals.

- Have your gas heater and appliances serviced regularly.

- Use less natural gas and electricity. Drive less and buy locally.

- Car pool, ride public transit, or walk or cycle whenever you can.

- Combine errands and shopping all in one trip. Resist the urge to buy an SUV and look for a smaller, efficient, lower-emissions vehicle.

- Keep your car tuned and smog checked. Replace the air filter regularly.

- Keep tires properly inflated.

- When driving, accelerate gradually, use cruise control on the highway, and obey the speed limit for optimum fuel consumption.

Be the Change You Want to See in the World

January 26 - Shop Green

- Buy items with less packaging, often larger sizes.
- Reuse water bottles.
- Buy items that can be reused.
- Take your own commuter mug to coffeehouses.
- Use rechargeable batteries in flashlights, toys, radios, CD players, etc.
- Use cloth diapers instead of disposable ones.
- Stop using disposable plates and cups.

January 27 - Bless the Beasts and the Children

While working to prevent waste and solve pollution, we are also indirectly helping to save the planet's animals, many of which are hard pressed to find a secure place to live and feed.

Defenders of Wildlife is . . . "dedicated to the protection of all native wild animals and plants in their natural communities." They do a brilliant blend of education and advocacy and are not afraid to go directly to the source to save nature. You can join for just $15. Log onto *www.defenders.org* and join the Defenders' Environmental Network (DEN) to receive timely e-updates on how you can help protect wolves, dolphins, bears, birds, endangered species, and key wildlife habitats, such as refuges and forests.

January 28 - A Juicy Life

Now is a great time to include citrus fruits in those six fruits and veggies you eat each day. Choose navel oranges, a huge pink grapefruit, squeeze fresh lemons for lemonade, limes for salsa and in margaritas, or grab a handful of kumquats to eat with lunch. If you have a juicer, buy a big bagful of organic oranges for your breakfast juice, make lemon juice ice cubes to pop into your gin and tonic or club soda. Get messy and peel a juicy grapefruit to share with your partner.

January 29 - Awakening the Senses

In the middle of winter, our senses can become muted. In the rain and snow, we don't look as far into the distance; without the green and blooming things of spring and summer, we tend to smell fewer heavenly smells. One perfect way to stimulate the senses is with aromatherapy, the use of particular essential oils to trigger and awaken our senses. Aromatherapy dates back to the Egyptians, Ancient China, the Greeks, and to Roman Times.

Choose oils that you love the aroma of. Remember, some oils are relaxing and some are uplifting and stimulating. Each oil targets a certain organ or area of the body but they are not medicines for a miracle cure and should not be taken internally.

The best way to apply the oils is in a massage, either neck and shoulders or the feet, where all the nerves of the body congregate. Or add 6-8 drops to a bathtub of warm water to relax in. You can also buy special diffusers to envelop the room with your favorite natural scent.

Be the Change You Want to See in the World

January 30 - One by One, We Can Do It!

Today or this week, try to do some jobs that you've been meaning to do around the house. You don't have to wait for spring cleaning fever to hit to get started.

If it'll give you the necessary motivation, go out and buy some new cleaning products. I use some great, earth-friendly products from Ecos. They're all biodegradable, natural, nontoxic, and not tested on animals.

January 31 - Breathe Easier

If you are a smoker, please consider quitting. Today. You don't need a lecture from me on all that's wrong with smoking. Just do it. Now.

FEBRUARY

Kindness Does Matter

February 1 - You Have a Dream, Too

Dream big, set goals, and don't let anyone stand in your way of being the best athlete you can be!

— Geena Davis

Today is "National Girls and Women in Sports Day." Started in 1987, it brings attention to the positive influence that sports participation has and how it advances equality.

The Women's Sports Foundation encourages moms to get more involved with their daughters' sports. Their Web site, at *www.womenssportsfoundation.org*, gives lots of advice on how to get involved, prevent discrimination, and increase participation, along with information on coaching issues, clinics, funding, and more. Check it out.

Sports can be a huge help with young girls' self-confidence with skill, determination, and inner success, a confidence that can carry through into her later years; it is said that if you succeed in sports, you lead in life.

February 2 - February Means Purify

This month derived its name from the Roman word *februare*, which means "to purify," since the festival of expiation (to make amends for any guilt or wrong doing) called "Februa" was celebrated at this chilly time each year. Do something today to purify your life.

February 3 - Eat Spicy!

At this time of year, especially if you have trouble with your sinuses, you might want to try some hearty, spicy meals, suitable for the cold season. Colds and flu can be relieved with warming spices such as cayenne, chili, curry, and ginger. Use in clear, steamy soups to clear your head, increase the circulation, and flush toxins. Increase your body's own immune system with foods high in vitamin C and beta carotene, foods such as garlic, onions, carrots, and broccoli. Ginger and garlic will help fight germs with their antibacterial properties and can help relieve sinus headaches and congestion by opening up those nasal passages and reducing swelling.

February 4 - Winter Wellness

Concentrate on "Yang" foods at this time of year: root vegetables, beans, and proteins.

February 5 - A Bowl of Yang

Gingered Vegetables and Rice

½	cup frozen peas
½	cup diced carrots
½	cup frozen corn
½	cup diced fresh tomatoes
¼	cup green onions, thinly sliced
3	cups cooked brown rice

Ginger Dressing

1	inch fresh ginger root, peeled and grated
1	tablespoon tamari or soy sauce
1	tablespoon toasted sesame seeds
¼	cup springwater

1	teaspoon sesame oil
2	teaspoons lemon juice

Steam peas, carrots, and corn until tender. In a medium bowl, stir together the tomatoes, green onions, and rice, and add the peas, carrots, and corn. In a small bowl, whisk together the dressing ingredients until well blended. Pour the dressing over the veggies and toss. Check seasoning. Serve warm on its own for lunch, or as a side dish. Serves 2.

February 6 - Savory Winter Stews

Adzuki Bean Stew and Dumplings

2	tablespoons olive oil
1	parsnip, peeled and diced
1	small yellow onion, peeled and chopped
2-3	cloves garlic, minced
2	carrots, diced
1	celery stalk, diced
2	potatoes, scrubbed and diced
3	cups boiling water
1 or 2	veggie bouillon cubes
2	bay leaves
1	tablespoon dried oregano
2	cups cooked adzuki beans
1	tablespoon tamari or soy sauce
1	tablespoon each fresh thyme, rosemary, and parsley

Be the Change You Want to See in the World

1   cup frozen peas

2   tablespoons cornstarch stirred with 2 tablespoons water

       freshly ground black pepper, to taste

Dumplings

1   cup whole wheat flour

2   teaspoons baking powder

½   teaspoon salt

¼   cup chopped fresh herbs such as parsley, thyme, etc.

½   cup plain soymilk

2   tablespoons olive oil

In a large pot over medium heat, sauté the parsnip, onion, garlic, carrots, celery, and potatoes in olive oil for about 10 minutes. Stir in the water, bouillon cubes, bay leaves, oregano, beans, and tamari. Cover and simmer for 15 minutes. Add more boiling water if it needs it. Stir in the fresh herbs, peas, and cornstarch paste. Season well. Simmer for a few minutes to thicken while you make the dumplings by sifting together the flour, baking powder, and salt, and then stirring in the herbs. Mix the milk and oil together, and then stir into the dry ingredients until moistened and combined. Drop by tablespoons into the stew. Cover; do not remove the lid for 15-20 minutes . . . no peeking now! Serve in bowls. Serves 4.

February 7 - Animal Testing Awareness

The perfume industry's methods of production are threatening the musk deer with extinction. Musk is found in male deer and has been treasured for centuries. Although today most perfumes using musk contain the synthetically produced kind, genuine musk is still used in some traditional perfume recipes, particularly in France.

The trade in wild plants and animals and their derivatives is big business, estimated to be worth billions of dollars and involving hundreds of millions of plants and animals every year. Be sure you know how animal friendly your perfume is.

February 8 - Diet For A Gentler Planet

"Farm" animals spend their entire lives crammed together in factory farms, dosed with pesticides and drugs, and living in fear and misery. Over 16 million are butchered every year to feed people in the U.S. Let's go back to happier eating, eh?

February 9 - The Peaceful Continent

Antarctica is the highest, windiest, and coldest continent and, inland, it is classed as a desert since the snowfall is so low. The ice sheet there contains 70 percent of the world's fresh water and 90 percent of the world's ice. No one lives there permanently since it is far too cold and hostile; however, scientists from twenty-seven countries work there under the Antarctic Treaty, the most successful international agreements ever signed. For details of this treaty go to www.antarctica.ac.uk.

This continent's weather plays a key role in the climate that effects the whole world and scientific research done here is vital in detecting climate change and ozone depletion. Any change in the ice sheet affects sea levels throughout the world and, since most of all major cities on our planet are situated along the coast, an increase in sea level could drastically flood these areas and potentially kill millions.

This ice sheet preserves a pristineness and a climate that spans a half million years . . . this will become lost forever if we allow it to become polluted or disturbed. Let's all help to preserve this remarkable continent and not ruin its beauty by human ignorance and greed . . . like we have the rest of our world.

February 10 - Retail Therapy

Giving shouldn't just be seasonal and charity does not begin at home, not when there are so many who need, really need, so much.

Check with your local social services and ask what you could buy for the elderly or for children. Then, armed with coupons and ads and specific ideas . . . shop 'til you drop! You could pick up winter coats, shoes, and sweaters for needy children—whatever is on your list, buy two, three, or four for the price of one. Then, remove the price tags, place in a pretty bag with colorful tissue paper, and hand deliver for smiles.

February 11 - Pure Bliss

Happiness is a perfume you cannot pour on others without getting a few drops on yourself.

—Anonymous

February 12 - Avoiding the Winter Blues

Many people are affected by the darkness of winter. But an extreme case of winter blues—a form of depression—is known as Seasonal Affective Disorder, or SAD, the symptoms of which are most pronounced in January and February, the months of least daylight in our Northern Hemisphere.

Research suggests that as many as 5 to 6 percent of adults in the U.S. may suffer from SAD. SAD's symptoms include anxious, empty, sad moods, changes in sleeping and eating habits—maybe craving starchy, sweeter foods. If it happened last year, and the year before that, then here's what you can do to help alleviate these feelings: spend as much time as you can out of doors, especially on sunny, bright days, and go for long walks to make you feel better.

If you have severe symptoms check it out with your doctor and get some help. There are specially designed light boxes that emit a very bright light through a filter in front of which the patient sits for a set number of minutes a day. Don't continue to suffer, get some help; it's not all just in your mind but a physiologically based form of depression.

February 13 - Midlife Self Care

So, how can we women control our menopausal symptoms without Hormone Replacement Therapy?

Eat a low-fat, plant-based diet. Dr. Dean Ornish and his colleagues at the nonprofit Preventive Medicine Research Institute have proven that changing to a low-fat, vegetarian diet, taking regular exercise, stopping cigarette smoking, and reducing your stress levels is powerful enough to actually *reverse* heart disease.

Do regular strength-training and weight-bearing exercises to fight osteoporosis. Resistance placed upon the skeleton during physical activity makes bones denser and stronger.

Reduce calcium loss (which leads to osteoporosis) by reducing your salt and caffeine intake. Also stop smoking and eliminate animal proteins from the diet (which will cut urinary calcium loss in half).

Eat soy foods. Soy foods are high in isoflavones—a natural plant estrogen which binds with our body's own estrogen receptors, lessening the highs and lows of mood swings. Research recommends 25 grams of soy protein and 40 milligrams of isoflavones daily. So choose fresh or frozen green soybeans (also called edamame beans), soy nuts and soy butter, or choose a soymilk you enjoy and drink two glasses a day. My favorite has 35mg of isoflavones in each eight-ounce glass and comes in plain, vanilla, coffee, chocolate, and strawberry flavors. Besides helping to regulate estrogen, soy can also help with other conditions such as osteoporosis, heart attack, and stroke. Women have an increased risk for these disorders during and after menopause.

Search out which perimenopausal products work for you. Some women swear by Black Cohosh and vitamin E for their symptoms. Black Cohosh is a plant estrogen herb that women have used for centuries to help regulate their hormones. Or try Evening Primrose Oil, Red Clover, and/or Dong Quai. Each woman reacts differently, so try them out yourself for 6-8 weeks to see if you get positive results.

February 14 - A Day Devoted to Love

On Valentine's Day each year, Americans buy more than a hundred million roses—three quarters of those are red, with pink coming in second.

Buy your share! Scatter coral petals in your bed and fill shallow bowls with red or pink and place them around the house. For that special, romantic meal, light unscented (scented candles can clash terribly with dinner aromas) red candles, get out the best china and crystal glasses, and fill a vase with water and a dozen red roses. Scatter more rose petals on the crisp, white tablecloth. Roses are our most enduring symbol of love—use them to excess.

At dinner, serve his most favorite vegetarian dinner to show you care, but make sure it isn't too heavy so that all he wants to do after is sleep!

February 15 - Chilly Days. Warm Heart

The best food is honest and simple, hearty and rustic, unpretentious and down-to-earth, and uses fresh, local ingredients. Even if you've never really gotten the hang of cooking, try a few simple dishes. Food doesn't have to be perfect. Just make it with love and the freshest ingredients and serve with a flourish. Set your table with plain pottery tableware and a loaf of crusty bread served on a wooden cutting board. Gather your closest friends and keep the whole scene comfortable and easy.

Be the Change You Want to See in the World

February 16 - Guilt-Free Comfort Food

Today try one of my favorite recipes for these chilly days.

Bean & Soysage Bake

2	tablespoons olive oil
1	large onion, chopped
2	packets Soysages, defrosted and cut into chunks
2	cans red kidney beans in vegetarian chili sauce
1¼	cups vegetable stock
4	teaspoons whole grain mustard
	hot sauce, salt and pepper, dried oregano, fresh parsley, and bay leaf, to taste
1½	pounds potatoes, thinly sliced
2	tablespoons of vegan margarine

Preheat oven to 350°F. Heat the oil in a 2-quart flameproof casserole dish, add the onion and Soysages, and sauté, stirring, until golden. Stir in the beans and sauce, stock, and half the mustard. Season and bring to a boil. Remove from heat and arrange potato slices on top, overlapping slightly, and dot with some margarine. Season the potatoes and then cover with foil or a lid and bake for 1 hour or more until the potatoes are tender. Remove the lid, dot the potatoes with more margarine and the remaining mustard, and return to oven to crisp, or broil until browned. Serves 4.

Down South Boozy Vegetable Bake

2	cups scrubbed, thinly-sliced potatoes (about ¾ pound)
2	cups peeled, thinly-sliced sweet potatoes
1½	cups peeled, thinly-sliced parsnip (about ½ pound)
1	onion, thinly sliced
¼	cup dried cranberries
1	tablespoon olive oil
	salt and freshly ground black pepper, to taste
1¼	cups vegetable broth
¼	cup maple syrup
1	tablespoon fresh thyme leaves
3	tablespoons bourbon
2	tablespoons fresh lemon juice
1	tablespoon vegan margarine

Preheat oven to 375°F. Combine the potatoes, sweet potatoes, parsnips, onion, and cranberries in an 11x7 inch baking dish coated with olive oil. Season with salt and pepper. In a medium saucepan, combine the broth, maple syrup, thyme, bourbon, lemon juice, and margarine, and bring to a boil before pouring it over the veggies. Cover and bake for 20 minutes. Uncover and continue to bake for another 50 minutes until the potatoes are tender and it all smells yummy! This dish goes great with steamed broccoli or collard greens. Serves 4-6.

February 18 - Planetary Pet Care

Each year, over 6 million dogs and cats are taken into shelters across the country; some are claimed by their owners, some are adopted into new homes, but a large percentage are destroyed, even though they are healthy and totally adoptable. The sheer number outweighs the availability of good homes. This is tragic. Reduce the number of uncared for puppies and kittens. The best way to do this is to spay and neuter— a simple surgical procedure that prevents animals from reproducing by removing their reproductive organs. Veterinarians perform spay/neuter surgeries under a general anesthetic and on animals as young as eight weeks old. The animal usually returns to normal activity within 24-72 hours, and any discomfort is minimal and well outweighs the suffering and death caused by uncurbed breeding.

February 19 - Consider the Future

One female cat and her offspring can produce over 420,000 cats in just seven years. How many will live in happy homes until they die? One female dog and her offspring can produce 67,000 dogs in six years. How many will receive a needle in the thigh before they are one year old?

February 20 - Respect Your Body

If you have a weight problem, check out *www.obesity.org*. It's a great Web site by the American Obesity Association—people who are speaking out for people with this disease and addressing obesity as a public health concern. Americans have increased their calorie intake per day by 163 calories and are heavier on average since 1980 by about fifteen pounds. We now live in an environment that almost seems to encourage people to become overweight; you don't even have to get out of your car to buy very high fat, calorie-rich, extremely cheap fast food, and it seems like every ad on TV is brainwashing you to eat this kind of food, which is very harmful to your health.

Motivate yourself by walking, eating fruits and vegetables, and banning fast food from your diet. Find a way to lose weight. Your life may depend on it.

February 21 - Tea with a Best Friend

Afternoon tea is usually served in England between three and four to help with that "sinking feeling" between lunch and a late dinner. It consists of a pot of tea served with milk and sugar, small savory sandwiches (crusts removed), scones with jam and cream, small cakes or slices of cake, and cookies.

Plan a tea party. It doesn't have to be stuffy and formal, just a great way for friends to meet and chat. If you have a tea shop nearby you could invite your friends to meet you there. Allow at least an hour so you're sure not to rush, or go once to get the idea, then invite guests to your own home. Arrange the furniture so that you can be cozy and intimate with a central coffee table for everyone to place her cup and saucer on. Keep everything small and dainty, from the crockery and spoons to the cloth napkins, cake stand, tray, and the food.

Start with tea to get the conversation flowing, and then serve savory foods, which you made beforehand and covered with a damp dish towel. Remove the crusts of the sandwiches of two or three varieties and cut to make four tiny ones—triangles or fingers are traditional. Allow four to six per person. Also serve small scones with either margarine and/or jam. Serve small cakes or slices of cake on another plate, or a tiered server is great, if you have one.

Pour the tea at intervals. Don't forget the strainer to prevent the leaves going into the cup, and make a fresh pot as needed. Use fresh water in the kettle each time, as water gets stale. I love to use pretty cups and saucers and tea plates; they do not need to match but should complement each other. Maybe you could choose a theme, such as roses, or a common color. Keep the sugar bowl and milk jug within this theme too.

February 22 - Cucumber Sandwiches and Tea Cakes

Yesterday we talked about having a tea party. Here are some of the foods I love to serve with my favorite teas.

Sandwiches: cucumber and fresh dill with margarine on white bread, vegan mock egg and watercress with vegan mayo on whole wheat bread, and vegan cream cheese with sliced tomato and basil leaves on white or wheat.

Cake: I like carrot cake, cut thinly.

Scones: plain, with or without currants, or cheese and herb.

February 23 - Bake Up Some Love

Cheese and Herb Scones

	light vegetable oil such as canola to grease the pan
2	cups self-rising flour
1	teaspoon baking powder
	pinch salt
2	ounces vegan margarine, cut into small pieces
4	ounces grated vegan cheese
1/4	cup fresh herbs (thyme, parsley, sage, chives, or oregano), finely chopped
2/3	cup plain soymilk

Lightly oil a cookie sheet or line with baking parchment. Preheat oven to 425°F. Sift the flour, baking powder, and salt into a mixing bowl. Add the margarine and mix with your fingertips until the mixture resembles breadcrumbs. Stir in the cheese, herbs, and milk to form a soft dough. Knead the dough very gently on a lightly floured board, and then flatten it out with your hand or a rolling pin to about 1 inch thickness. Use a small biscuit cutter to cut out about a dozen or so scones. Brush the tops with a little soymilk. Place in the hot oven for 10-15 minutes until golden brown. Transfer to a wire rack to cool a little. Serve on the same day sliced horizontally with a little margarine. Serves 12.

February 24 - More Beef = Fewer Trees

The next time you consider grabbing a burger at a fast-food place, remember this: over the past few decades, the rainforests have been disappearing to satisfy our "hunger" for cheap beef.

Rainforests are home to over a thousand indigenous tribal groups, thousands of species of birds and butterflies and exotic animals—all of which are now endangered. Rainforests also affect rainfall and wind all around the world by absorbing solar energy for the circulation of our atmosphere: the trees provide buffers against wind damage and soil erosion, which then helps prevent flooding along our coastlines. They are a precious part of our ecosystem. Let's all do something to protect them.

Over five million acres of South and Central American rainforest are cleared each year for cattle to graze on. The local people don't eat this meat—it is exported to make the $1 hamburger and a cheap barbeque meal. The most important thing you can do is to encourage everyone to stop eating meat, or at least cut down.

February 25 - Recycling Saves the Rainforest

Here are some other things you can do to save the rainforest.

• Don't keep tropical birds or reptiles as pets. Let them live in nature.

• Buy items made from sustainable wood. Hardwood teak and rosewood encourage logging and deforestation, another rainforest destroyer.

• Recycle all your cans. Bauxite is mined from the ground in tropical countries and is the source for aluminum.

• Buy local, organic food whenever possible. Conventional agriculture is exhausting our forests' resources.

• Support any organization that is legitimately working to protect the environment in developing countries and in precious rainforests.

February 26 - Proud to Be Green

Visit www.therainforestsite.com or put the words "rainforest" into your search engine and educate yourself, your family, and your friends further. Be proud to be green.

February 27 - Eating With a Conscience

The western world hasn't always eaten so much meat. As Les Inglis, in his brilliantly informative book, *Diet For a Gentler World: Eating with Conscience*, informs us:

> A hundred years ago, an eighty-acre farm in Missouri provided a decent living for a family of five. Tomatoes, potatoes, corn, squash, and other garden vegetables grew in the fields, and trees near the house provided peaches, apples, cherries, and plums. The animals on the farm provided labor for the most part, not meat. Besides two horses and a mule, a cow was kept for milk and only occasionally a pig for meat. A yard of chickens was the main source of the family's animal protein.
>
> At midday the farm wife laid out the largest meal of the day, occasionally featuring meat but always potatoes, at least two vegetables, breads, or biscuits, fruit preserves, honey, and a dessert. This typical American family of days gone by depended far less on meat than does today's typical American family, which produces nothing it eats.

No one is saying their life was ideal; we know it was very hard work, but the food they ate was healthier and more nutritious. It was organic, free of hormones, the chickens happily scratched around in the dirt, providing free-range eggs, and the cows ate grass and wandered the field, contentedly chewing the cud. Doesn't that sound better for us all?

February 28 - Consider the Climate

Climate change is a very real threat to us all. Even though it is the dead of winter, consider one way of cutting your contribution to global pollution and do it today!

MARCH

3

Celebrate All Life

March 1 - Make Time in Your Calendar for Health

Spring is fast approaching and it's a good time to be thinking about new beginnings. How about trying to find a little balance in your life? A little planning can really help. Put aside some time each day to plan the following day. If you like lists, write down the things that need to be done, organize your day and you'll feel less stressed. Check your calendar for any appointments, such as the dentist or having the car serviced.

Only do the things that really need to be done: leave the vacuuming and the polishing for another day if you don't have time. Do larger laundry loads a few times a week rather than small ones each day. Cook enough food for today and tomorrow so that you only have to clean the kitchen once—if you serve it with a different carbohydrate it will seem like a whole new meal. For example, serve a vegetable chili with rice the first day and crusty bread or a baked potato the second.

Making compromises in your schedule can prevent you from compromising your health. Be sure you find time each day to do those things that are important to you and your well-being, whether it be a craft, reading, going for a walk, meditation, or sitting with your loved ones.

March 2 - A Vegan Diet Equals a Reverence for All Life

A respect and reverence for all life means not buying anything leather—shoes, belts, purses, dog leashes, car seats, jackets, gloves, luggage, or furniture. Next time you go to buy something leather, see if there is a pleather alternative first.

March 3 - Celebrate Mardi Gras

Mardi Gras, or "Fat Tuesday," falls 47 days before Easter Sunday, any time between February 3 and March 9. Early French explorers celebrated on the banks of the Mississippi as long ago as 1699 and New Orleans has since added parades and extravagant masked balls.

In honor of New Orleans and all it has been through let's make this day our Mardi Gras. If you can't *laissez les bons temps rouler* in New Orleans, invite some friends 'round and deck yourselves, your house, the table, and the food in gold for power, green for faith, and purple for justice—and have some fun.

Serve food with a Creole flair flavored with Cajun spices, such as a red bean dip with potato chips or garlic mushrooms with French bread. For entrees think spicy rice-stuffed red peppers, mashed sweet potatoes, cornbread, vegan jambalaya, and Louisiana red beans and rice. Drinks? How about mint juleps, bourbon, Planters' Punch (1 oz. dark rum, 3 oz. pineapple juice, ¼ oz. grenadine, 2 oz. orange juice, served over ice. You can leave out the rum for a delicious nonalcoholic version).

Add your fun-loving friends and let your Big Easy Bistro roll.

March 4 - Fat Free Fat Tuesdays!
Creole Mardi Gras Jambalaya

4	ounces firm tofu, diced
2	tablespoons olive oil
1	onion, chopped
2	bay leaves

1	cup uncooked white rice
1	teaspoon salt
12	teaspoons Cajun Creole seasoning
2	teaspoons tomato paste
3	cups boiling water, or more
1	sweet potato, peeled and small cubed
1	carrot, chopped
1	cup frozen corn, defrosted
1	can (15 ounces) red kidney beans, drained and rinsed

Dry sauté tofu in a large, nonstick fry pan until golden. Remove and set aside. Add the olive oil to the pan and sauté the onion and bay leaves for a few minutes. Stir in the rice, salt, Cajun seasoning, and tomato paste. Stir in the water and bring to a boil. Then stir in the potato, carrot, corn, and beans, as well as the reserved tofu. Cover and cook on a low heat until the rice and sweet potatoes are cooked, 15-30 minutes. Taste and add more seasoning if required. Add more boiling water during the cooking if needed. Serves 4.

March 5 - Rice Is the Grain That Feeds the World

My favorite is Basmati, which means "one with a good smell" and originated in India and the foothills of the Himalayas. Brown basmati still has the bran which contains B vitamins and fiber, but it takes longer to cook since the bran is a barrier to the cooking water. Rinse basmati in a bowl of water and change the water a few times until clear, since the dust, cornstarch, and other impurities found on the rice do not feature in my recipes!

Be the Change You Want to See in the World

Rice is the grain that feeds more people in the world than any other and dates back many thousands of years—to 6,000 B.C. in northern Thailand, 5,000 B.C. in Northern India, and 10,000 B.C. in Kashmir. It is grown in paddy fields (water-filled beds) in countries all over the world—India, China, Indonesia, Thailand, Japan, the U.S., Greece, Turkey, the Mediterranean, and others.

There are so many wonderful ways of using rice that you could cook one per day for the rest of your life and not run out. Just be sure to use the correct rice type for the recipe or the dish will not be successful. For example, risottos need Arborio rice, a fat, starchy grain, which cooks to a soft, creamy texture yet retains a firm middle. Jasmine is an aromatic rice and used in Thai cooking: it is a long-grained yet sticky rice that remains moist and tender.

Don't be afraid of rice. Just learn how to cook each type, use a heavy pan, and don't overcook—or buy a rice cooker. The following two days include some of my favorite recipes.

March 6 - True Health is Where East Meets West

East Meets West Rice Salad

Rice

2	cups water
1	teaspoon salt
2	cups uncooked sushi rice, rinsed well and drained

Dressing

1	tablespoon sunflower oil
1	tablespoon sesame oil
1	tablespoon soy sauce or tamari
½	cup rice vinegar
1	teaspoon grated fresh ginger
¼-1	teaspoon wasabi (Japanese horseradish), optional

Other Ingredients

¼	cup finely chopped red onion
1	cup English cucumber, cut into strips or diced
1	tablespoon toasted sesame seeds
1	sheet nori seaweed, cut into narrow strips

Bring the water to a boil in a medium pan. Stir in the salt and rinsed rice. Reduce heat, cover, and simmer for 20 minutes or until cooked and the water has been absorbed. Combine the dressing ingredients in a small bowl and whisk well. In a large serving bowl, combine the rice, dressing, onion, cucumber, and sesame seeds. Sprinkle the nori over the top. Serve warm or cold. Serves 4-6.

March 7 - Aloha in a Bowl

Waikiki Rice Bowl

2	cups white basmati or jasmine rice, rinsed
1	cup water
1½	cups coconut milk
1	tablespoon canola oil
1	teaspoon ground cumin

½	teaspoon cinnamon
1	small red chili, seeded and finely chopped
1	carrot, cut into julienne strips
1	stalk lemongrass, cut into 2 inch pieces
½	teaspoon salt
1	cup frozen peas
	toasted coconut

Place the rice in a large bowl and pour the water and coconut milk over it. Stir and set aside. In a large skillet, heat the oil and sauté the cumin, cinnamon, chili, and carrot over medium-high heat for a minute. Stir in the lemongrass, salt, and rice (along with its soaking liquid). Bring to a boil, cover pan, reduce heat, and simmer for 15 minutes. Stir in the peas, cover, and simmer another 5 minutes. Remove the lemongrass and garnish with toasted coconut. Great with a broccoli stir-fry or a spinach salad. Serves 4.

March 8 - Companion Pets

If you or someone you know is thinking of adding a companion animal to the family, here are some things to consider before adopting from a shelter or other rescue group. Companion animals are not toys that you give attention to for a few minutes each day and then banish to the backyard. Nor are they to be returned when the novelty has worn off or they shed hair on your best sofa.

• The life of a companion animal will vary between five and fifteen years and you should allow for that time in your planning.

• Allow the animal time to understand what you want of them, time to trust you, and time to learn to love you.

- Never be angry with an animal. That is frightening and they do not understand. Never hit an animal, lock them up. or rub their nose in an accident—these are human ways not understood by a furry one.

- Take extra care of an animal when it gets old; you too will be old one day.

- Go with your friend on its last difficult journey; it is easier for them if you are there.

March 9 - Vegetarianism Fights World Hunger

There is an ever-worsening food crisis in many countries in Africa and other parts of the world. It's true that a lot of people are starving due to wars, corrupt politicians and greedy dictators, but enough food is produced worldwide. It just doesn't reach the hungry.

Another part of the problem is that 90 percent of U.S. and European grain goes to feeding cattle and other livestock. Now if more people would become herbivores again and we sent more excess grain directly to famine areas, the hungry could become stronger, overthrow their corrupt regime by peaceful force with women in the main negotiating positions, and these Third World countries, who are more than capable of feeding themselves, could do so again, and they could prosper and thrive.

March 10 - Self-Care at Work

When at work, try not to get so wrapped up in what you are doing that you forget about yourself.

Be sure and take your allotted lunch break away from your desk. Go out into the fresh air, take a walk, look up at the clouds, down at the grass and flowers, relax your shoulders and relax your mind.

Choose low-fat food for lunch salads, fruit, light sandwiches, so that your body won't be left sluggish for the afternoon, and enjoy every bite. Eat with friends and enjoy their company and have fun, or eat alone and have a quiet break to relax and unwind, ready for the afternoon.

March 11 - Self-Care for Women

Invasive cervical cancer could be as rare as polio is now if every woman had a routine pap smear. In fact, cervical cancer used to be the number one cancer killer of women, but since the simple lab test created by George Papanicolaou came into being, deaths have decreased by a huge 70 percent over the past 50 years.

During a Pap smear, cells are collected from the cervix—a fast, painless procedure. These are then sent to a lab, viewed under a microscope, and deemed normal or abnormal. Every woman who is sexually active should have a regular Pap smear. As women get older the risk of cervical cancer increases.

For the most accurate results the ideal time to have one is two weeks after the first day of your last menstrual cycle. Don't use vaginal creams or medications 72 hours prior and don't have sexual intercourse within 24 hours of your appointment.

Of course we all hope to hear, "It's normal!" but if they say, "Abnormal," don't panic. Most cases of abnormal Pap smear results do not indicate cancer, it just means you will have to be retested, and your doctor will discuss it with you. Remember that this test does not detect cancers of the ovaries, uterus, or fallopian tubes, only of the cervix.

Annual Pap smears are a great gift to give your health and your peace of mind.

March 12 - Walking for Health and Happiness

Exercise not only lifts your sagging butt, it even lifts your sagging spirits and brightens your mood, indeed your whole outlook on life. A ten to twenty minute walk each day may even save your life, being as beneficial as giving up smoking is in the fight against heart disease.

Invest in a pair of walking shoes and go for it. Walking is a great route to happiness and health. For your heart's sake, grab a friend or a family member or two, open the door and walk. It's easy!

March 13 - Organica

We've all seen the organic section at the market and we see the word on other produce, but what does it really mean, and is it worth paying the extra?

Organic is all about not using harmful insecticides—no herbicides, fungicides, and no chemical fertilizers where crops are grown. The area must also be free of chemicals for at least three years.

And the answer is yes!

March 14 - Nature's Way

The object of organic farming is to produce the highest quality of food, with the highest yield, but without ruining the environment and poisoning the soil. It's also best if the crops grown are appropriate to the area, both in climate and soil conditions, are suitable for locals' eating requirements, and that genetically modified seed is not used.

Buy organic whenever you can.

March 15 - What Really Matters

What lies behind us and what lies before us are tiny matters compared to what lies within us.

—Ralph Waldo Emerson

March 16 - Good Luck

The shamrock was a charm in pagan times against witches, fairies, and all sorcery. A four-leafed one, found accidentally, now promises good luck.

March 17 - A Celtic Celebration

Today is the feast day of the patron saint of Ireland, St. Patrick.

St. Patrick's Day is a good excuse to celebrate—even if you're not from the Emerald Isle. Invite some friends round for a *ceili*—a time of traditional singing and dancing. Hang a banner with the words *Cead Mile Failte*—100,000 Welcomes. If your guests can't sing get them to recite a poem, read an extract from a play, or tell a tale full of Irish blarney! Choose a piece by an Irish person, such as Iris Murdoch, Maeve Binchy, Oscar Wilde, George Bernard Shaw, or poet William Butler Yeats.

Serve up hearty Irish vegetarian food—easy vegetable soups with potatoes and cabbage served with soda bread. Make shamrock-shaped cookies with green frosting, or green tea ice cream. Drink Guinness, green crème-de-menthe, and Irish coffee. Wearing and eating something green on St. Paddy's Day is said to bring good luck.

March 18 - Promises of Luck

Take a day and celebrate yourself.

March 19 - Go Ahead and Be Great!

It is never too late to be what you might have been.

—George Eliot

Are you living your life to its full potential, or do you feel angry and frustrated that your true worth has not yet been discovered? Do you even know what your life's calling is? You have to have the dream; no one can have it for you. Then work towards that dream. Just knowing that you were put on the Earth to be a great writer, a nuclear physicist, or a leader for world peace won't cut it, however. A dream alone will not make you anything but a dreamer. You have to start at the bottom and put in the groundwork and finally, if you are lucky and have worked hard enough, you may get there. Success is in the effort.

Be the Change You Want to See in the World

March 20 - The Nonviolent Diet

Each person who gives up eating meat and dairy can save 37 animals . . . each year. Over a lifetime that will save 2,700 animals.

March 21 - Compassionate Food Choices

Go to www.meatout.org, or any vegetarian Web site, to learn more about your new, healthy way of eating. It simply is not humane to raise and eat farm animals when there are other healthier and more compassionate food choices available.

March 22 - Olive Oil - The Essence of Civilization

Unlike women, olive oil does not improve with age! Olive oil must be used within the first year of pressing, and for the best flavor within three months of opening.

This was told to me by a very elderly Cretan lady (actually her granddaughter translated for her) who was sitting under her olive trees in a village we were visiting in the Crete countryside. I was buying some of the cold-pressed, green liquid to take home and was invited to sit in the shade for a while.

Experiment with olive oils yourself to find those you like the best. Some oils are best for cooking, others for salad dressing or drizzled over steamed vegetables, others for dipping crusty bread in . . . it all depends on the oil's acidity—some are fruity, others sweeter, or peppery, or green.

March 23 - Crete - Veggie Style

Here are some recipes I brought back from Crete.

Greek Lentil Salad

¾   cup dried green lentils, rinsed
3½   cups water
¾   cup bulgur (cracked wheat)
1   organic tomato, chopped
1   small cucumber, diced
¼   red onion, chopped
1   cup pitted black olives, sliced

Greek Vinaigrette

1/3   cup virgin olive oil
3   tablespoons red wine vinegar
1   tablespoon lemon juice
1   teaspoon Dijon-style mustard
1   tablespoon fresh oregano, chopped
1/2   cup vegan feta cheese, to garnish

In a saucepan, mix the lentils and 2 cups of water and bring to a boil. Reduce the heat, cover, and simmer for 20 minutes or until the lentils are cooked. Drain and rinse. Meanwhile, in another pan, bring 1½ cups water to a boil, stir in the bulgur, cover, and turn off the heat, leaving the bulgar to soak for 30 minutes. Drain away any excess water. In a large serving bowl, toss the cooked lentils, bulgur, tomato, cucumber, onion, and olives. In a screw topped jar, shake together the vinaigrette ingredients and tip over the salad. -Toss gently but well. Cover and refrigerate for 4-8 hours. Before serving, top with feta. Serves 4.

March 24 - Off the Vine

Greek Dolmades

4	cups water
1	cup long-grain brown rice
½	teaspoon salt
1	tablespoon olive oil
1	small onion, finely chopped
1	clove garlic, crushed
2	tablespoons fresh Italian parsley, finely chopped
2	tablespoons pine nuts, toasted
2	tablespoons dried apricots, chopped
	grated rind of 1 lemon
1	tablespoon fresh lemon juice
	salt and freshly ground black pepper, to taste
1	can (14.5 ounces) crushed tomatoes, undrained
24	bottled large grape leaves (vine leaves)

Bring the water to a boil and stir in the rice and salt. Reduce heat, cover, and simmer for 30-40 minutes or until the rice is cooked and the water absorbed. Cool. Heat the oil in a large skillet over a low heat and stir in the onion and garlic. Cook for about 10 minutes until softened. Put aside about ¼ cup of this to add to the tomatoes later. Stir the remainder into the rice, plus the parsley, pine nuts, apricots, lemon rind, and lemon juice and season well to taste. In a large skillet (you will be adding the rolled-up grape leaves into this later) stir the tomatoes and reserved onion together and simmer for 10 minutes. Turn off the heat and season. Rinse the grape leaves with cold water and pat dry. Remove stems and throw away. Spoon 1 rounded tablespoon of the rice mixture into the center of each vine leaf, then bring the opposite points of the leaf up and fold over the filling. Then roll the leaf up tightly and place, seam side down, in the tomato sauce. Repeat with all the leaves, snuggling them all together in the tomato sauce. Cover the pan and simmer for 20 minutes until heated through. Serves 4-6.

March 25 - Animal Rescue Angels

When I visited Shambala, an animal refuge near L.A. founded by actress Tippi Hedren, she told us there are more tigers kept captive in Texas than are wild in India. The animals usually arrive at places like Shambala underweight, sick, sometimes vicious and unapproachable, and nearly always depressed. At these refuges, they receive love and kindness, people-grade food, the best vet care, the company of their own kind, and are regularly moved from one huge grassy area to another by special fenced corridors for a change of scenery and to prevent boredom—what a concept!

Click on *www.shambala.org*, or search out your own local wildlife refuge. Visit and support one in whatever ways you can. See rescued cats like Tamara, a tiger cub who was being sold out of the back of a station wagon at a Southern Californian shopping mall, or Kara, a black leopard abandoned in sub-zero temperatures in a garage in Wyoming with no food or water and suffering from frostbite on her ears, paws, and tail. These animals belong in the wild . . . not in a backyard or basement.

March 26 - Opting Easily Over Eggs

Many people choose not to use eggs in their diet. Eggs may be considered nutritious but they also carry salmonella (which is the leading cause of food poisoning in this country), contain saturated fat and 213 milligrams of cholesterol per average-sized egg, and many people are allergic to them. All the nutrients in an egg can easily be obtained in a vegan diet, but without the health risks and terrible cruelty involved in their production.

Look for eggless egg replacer powder products in your store and use as the package instructs, or for baking use one tablespoon of mashed tofu in place of an egg, or a tablespoon of cornstarch or soy flour plus two tablespoons of water. As a binding agent in veggie burgers or loaves substitute tomato paste, moistened breadcrumbs, oats, or mashed potatoes for the eggs.

March 27 - Dog Only Knows

The average dog is a nicer person than the average person.

—Andy Rooney

March 28 - Dog Forbid

An untrained dog is a misery to live with and is one of the main reasons why millions of dogs across the country are abandoned by their owners and are ultimately euthanized. Dogs are like small children and have to be reminded what is and isn't acceptable behavior—rules and regulations set by you and carried out in a calm, respectful manner with consistency and patience.

A disobedient dog is like a naughty child—he does it because he can, because you let him get away with it. Set rules and boundaries and you will have a happy companion animal, an asset to you, and a friend for life.

March 29 - Only Connect

As we come out of the long winter, make an effort to reconnect with the world.

March 30 - Finding New Ways to Live

Basically we should stop doing those things that are destructive to the environment, other creatures, and ourselves and figure out new ways of existing.

—**Moby**

March 31 - Moving Through Life

In ancient Sparta, a very advanced culture, women were required by law to be fit and to exercise. Throughout life all women face hurdles—there always seems to be some sort of barrier trying to stop us succeeding and being happy. Taking part in games helps you overcome them and find the strength to continue life. Exercise, team spirit, and being with other women helps develop friendships, communication skills, cooperation, and a feeling of belonging; it helps depression and shyness, and develops a sense of confidence.

This spring, why not join a ladies' team of soccer, volleyball, basketball, softball, or another athletics team. Ask a friend if she'd like to go, too. Not that energetic or competitive? Try it. Start small and develop from there. Have fun, but get out there.

APRIL

Nature Renewed

April 1 - Laughing is Good For You

The gods too are fond of a joke.

—**Aristotle**

Happy April Fools Day! We can shed winter's despondency at last. April is the perfect month to clean out the closet, take our exercise outdoors, reconnect with our spiritual side, celebrate Earth Day—and today, make a date with your funny bone. If you like sending unsuspecting people on errands for a left-handed hammer, skyhooks, or tartan paint then go for it! Just be sure that whatever jokes you do play are harmless and fun.

April 2 - We Are Also Our Ecosystem

I used to climb trees and watch butterflies and emerging tadpoles swim in the river. Now natural history is taught less and less in our schools. We are less aware of what is happening in our countryside, our eyes glued to the TV and our computer screens. Will future generations even know their native species have gone, that a local wooded area has been cleared for a shopping mall, a highway now follows a course where a river used to be, and the creeks, ponds, and streams have dried up and taken their wildlife with them? Every ecosystem matters and we must take a stand to save it, we must continue to indulge our love of wildlife, we must take action . . . before it all is cast in concrete. Find a local cause in your area and give it your support, either with money or your voice.

April 3 - Joy is Natural; Nature is Joy

Joys come from simple and natural things; mist over meadows, sunlight on leaves, the path of the moon over water. Even rain and wind and stormy clouds bring joy.

—**Sigurd F. Olson** (1899-1982), American naturalist, *Open Horizons*

April 4 - It is Never Too Late to Begin Getting Fit

Back in my days of managing a health club in London I had members who came in to weight train. They could pump iron with the best of them—heavy weights, then pose and flex in the mirror. Other members would be envious and confide to me they wish they were as fit. Actually, fitness isn't just about strength. Those guys rarely did any cardio work to keep their hearts and lungs strong, they did no stretching, and their diet was far from healthy—too much protein and cholesterol and certainly not six fruits and veggies each day.

To be fit means to take your body through all aspects of cardio-vascular, strength, flexibility, agility and balance, to eat regularly and sensibly, to drink loads of water and get plenty of rest and sleep. The good thing is . . . it's never too late to start. Keeping fit helps you stay healthy, staves off osteoporosis and muscle wasting, keeps the heart strong, gives you more energy and the ability to face life's ups and downs by giving you that lift to make you look and feel a whole lot happier.

You don't have to lift huge weights and spend hours in a gym, you don't have to be able to run a marathon, or be an Olympic gymnast, but you do have to cover all aspects of fitness and do them regularly. Do something every day. Set a schedule . . . and stick to it.

April 5 - National Soybean Month

April is National Soy Month and such a versatile food source is very easy to incorporate into everyone's diet, especially since the U.S. remains the world's top producer of the bean. Soy is low in fat, low in cholesterol, and high in isoflavones—a woman's best friend in the fight against breast cancer and hormone fluctuations.

Soy products are easy to use and prepare, are extremely versatile, very healthy, and inexpensive. Incorporate fresh beans into Japanese cooking, soups, and vegetable stews. Use soy sauce and tofu in Chinese stir-fries and use veggie burgers and Soysages on the barbeque.

Nongenetically modified soybeans are the food of our future—eat them often.

April 6 -Soy Joy

Originally, soy beans were grown in the U.S. in the late 1800s to be ploughed back into the ground as a fertilizer to nitrogen enrich the soil. Later, soybeans were used as cattle feed and the oil extracted to make soap. Today we realize how healthy, tasty, and versatile the soy crop is for humans: as a fresh vegetable called edamame, pressed to make bean curd called tofu, made into veggie burgers and other soy meat substitutes, soy bean oil, soy butter (like peanut butter), soymilk in many flavors, soy cheese, sauce, paste, fermented to make tempeh, or dried as a legume. This bean has been cultivated in China for over five thousand years and has been a staple throughout the East for its high protein, iron, calcium, and B vitamin content.

April 7 - Eat the Food of the Future, NOW

Italian Soy Balls

1	tablespoon olive oil
1	large yellow onion, diced
1	green bell pepper, seeded and diced
1	red bell pepper, seeded and diced
1	cup celery, diced
2	cloves garlic, crushed
1	bag vegan soy balls
1	jar of your favorite marinara sauce
2	cups frozen edamame (soy) beans
1	tablespoon dried oregano
1	teaspoon black pepper
1	cup fresh herbs, such as parsley, oregano, and/or chives
	Cooked white basmati rice or pasta to serve

Heat the oil in a large saucepan and sauté the onion, bell peppers, celery, garlic, and soy balls for about 6-8 minutes. Stir in the marinara sauce, beans, oregano, and black pepper. Either simmer over a low heat for 25-30 minutes, or tip into a preheated slow cooker and cook on low for about 6 hours. Check seasoning—add salt if you must!—and stir in fresh herbs. Allow to stand for 5-10 minutes before serving with rice or pasta. Serves 4.

April 8 - The Spices of Life

Malay Tofu Curry

1	package (12 ounces) firm tofu, drained and cubed
2	tablespoons canola oil or ground nut oil (if you have it)
2	inch piece fresh ginger, peeled and grated
2	red chilies, left whole
2	cardamom pods (optional)
2	cloves garlic, crushed
2	teaspoons garam masala
2	teaspoons ground cumin
2	teaspoons ground coriander
½	teaspoon turmeric
1	teaspoon salt
2	slices canned pineapple in natural juice, chopped
	juice from the pineapple can
1	cup edamame (soy) beans, defrosted if frozen
½	can coconut milk (shake well before pouring)
	jasmine or red rice, to serve

Sauté the tofu in the oil until browned a little, drain on kitchen towel, and set aside. In the same saucepan add the ginger, chilies, cardamom, and garlic and sauté for a few minutes. Stir in the garam masala, cumin, coriander, and turmeric and lightly fry for 2 more minutes. Add the salt, tofu, pineapple, pineapple juice, and edamame beans. Simmer for 10 minutes. Stir in the coconut milk. Heat through, check the seasoning, and serve with jasmine or red rice. Serves 4.

Be the Change You Want to See in the World

April 9 - Plan Your Garden and Your Life

Gardens, like life, need a certain amount of forward planning. This can be best achieved upon reflection, by looking at those photographs or by walking in the garden and seeing what works well. Knowing your soil and where the shady or sunny spots are will determine the success or failure of some plants; as will overwatering. Which parts of the garden were disappointing; why? Which parts successful? Build on that knowledge too. Take out your journal and make notes to consult later in the planting calendar.

Gardens are a continual work in progress, again, like life. We can learn by our mistakes, or we can ignore them. Your life and your garden should remain colorful and interesting, satisfying, and harmonious. So take stock, make plans and get your garden, and yourself, ready for spring's rejuvenating beauty.

April 10 - The Mindful Mom and Dad

I believe that most parents try very hard to be conscientious when it comes to leaving their children with sitters. Let me add a few safety reminders to your list of instructions.

• Show the sitter where the fire extinguisher is.

• Leave a flashlight handy in case of a power failure.

• Tell the neighbors you are going out in case of an emergency.

• Keep a first aid kit on hand and let your sitter know where it is.

• Be sure your sitter knows child CPR.

• If you have a baby, be sure the sitter also knows how to lie him or her back down after a wake up—with no stuffed toys or heavy blankets or other items that may choke.

All scary stuff, but literally life and death.

April 11 - Seven Generations

In our every deliberation we must consider the impact of our decisions on the next seven generations.

—from the *Great Law of the Iroquois Confederacy*

April 12 - Spring Clean Green

Most of us love the springtime, but I doubt our environment does: springtime means spring-cleaning. If you use all those chemicals you bought at the market and multiply that by the millions of homes across America, look at the harm we cause.

Most of the cleaning products used are petro-chemicals, manufactured from the oil industry, which are cheap to buy but expensive to our natural world and can be harmful to our health. Many cleaners, solvents, insecticides, and polishes contain ammonia, chlorine, and/or formaldehyde, which are extremely toxic and harmful to our families and pets. If it's stinging your eyes and you have to wear rubber gloves then it makes sense it's not healthy. These cleaners can cause lung, skin, and eye irritations and may also be carcinogenic.

It's time to make conscious decisions to keep ourselves and our environment safe by deliberately buying nontoxic, hypo-allergenic, and biodegradable products which use safer ingredients such as citrus oils, glycol-based degreasers, and are phosphate free. Support companies such as Planet, Ecover, Seventh Generation, Earth Friendly, and Heather's Natural Products: products that are as gentle on our planet as they are on us, with no harmful fumes. One of these brands should be available at your local market, so please support them. They may seem expensive at first glance but many are concentrated and anyway, what price is there for your children's health? (They are also not tested on animals so no more drip, drip, drip of liquid bleach in a precious rabbit's eyes.) For the sake of everything we hold dear, read labels and choose very carefully the household products you buy. And don't forget the virtues of old-fashioned elbow grease.

April 13 - Spring Cleaning for the Soul

Reward yourself for spring cleaning the house by spring cleaning yourself; take in warmer air and fresh breezes of flowers and mown grass and feel how much brighter and more alive you feel.

April 14 - Clean Skin, Radiant Skin

Gently exfoliate your skin before a bath or shower with a cactus-bristle brush: start at your toes and work up towards the heart, ending with the fingers and arms to get blood and lymph flowing. Then use your favorite not-tested-on-animals perfumed bath or shower products. Finish with delightful body and face lotions.

April 15 - Naturally Refreshing Scents

Buy some new, pretty bed linens and spray with rose or lavender water. Check your lingerie drawer and throw away those old, unflattering panties and bras. Add scented drawer liners and herbal sachets. Invest in new underwear to make yourself feel luxurious in lace and satin without sacrificing support and comfort. Visit the fragrance counter and choose something light and feminine and cruelty-free to give you a newfound confidence.

Feel good about yourself this spring—be uplifted, invigorated, relaxed—be a new you!

April 16 - Grow Your Own Salad

Did food taste better when you were a child? That's probably because food used to be grown for flavor and not for a uniform shape and size like today's fruits and vegetables. So, why not take your taste buds back to those halcyon days and grow your own favorites? What you choose may depend on how much room you have, but tomatoes are a good place to start.

Tomato plants can be bought at any garden center and should be suitable for your climate and situation. Most are easy to grow: hybrids like Better Boy, Early Girl, and Champion, or choose heirloom varieties, those that date back generations and have been handed down through the family.

Tomatoes need a slightly acidic soil of rich compost that isn't too high in nitrogen or you'll end up with huge plants but small fruits. Wait until your nighttime temperatures are over

50°F and choose a position of full sun for a minimum of six hours a day. Place a healthy looking plant (one that has a good color but isn't too leggy) into a nice deep hole, either directly into the ground, a raised bed, or a large pot or half-barrel, and don't forget to include some high-phosphorous plant food to encourage flowers and fruit. Plant deep, and it's okay to bury the lower shoots as more roots will sprout from their bases, but leave 3-4 branches showing from the base.

Tomatoes just love evenly moist, well-drained soil and don't forget to add a trellis when you plant to support the plant as it grows. Feed every two weeks with a special tomato fertilizer and, if you get aphids, blast them with a strong stream of hose water, spray with an organic insecticidal soap, or buy some ladybugs and encourage those onto the plant. I grow lavender in pots next to my tomato plants, which the bees love, and in return they kindly pollinate my tomato flowers.

You may need to experiment with what works in your garden, but nothing beats the taste of a freshly picked, still warm from the sun, bright tomato, whether it be a tiny cherry variety or a huge slicing one that is pale or deep red, yellow, orange, even black or striped. Have fun!

April 17 - Eat More, Weigh Less

Is your body ready for the summer? Don't even think about a crash diet—it'll leave you bad tempered, and stressed, and you know you'll put that weight back on in time. For long-term weight loss and general good health it makes sense to alter your whole life style: not just the amount of food on your plate, but what kind it is.

The food you eat should be nutritious and varied. Eat at least six pieces of fruit and vegetables each and every day; eat low fat foods, whole grains, and legumes.

To lose weight the average woman needs to consume 1,800 calories a day, eating less may well slow down your metabolism. Write in your journal everything you eat each day for say a week, to help you see what you can cut down on without even noticing: that candy bar you eat after lunch, change it for a piece of fruit; swap that second martini for a club soda with a twist of lime; cut out that second scoop of ice cream. Instead of snacking during the day, drink a large glass of water—it'll fill your stomach.

Don't think diet, just change your outlook.

April 18 - A Little Bit Every Day Adds Up Over Time

By cutting just 100 calories a day, every day for a year, you could lose ten pounds. That's not so bad, is it?

April 19 - Listening to Your Animal Friends

A growl is often a warning.

April 20 - Teach Your Children Well

A statistic on the Humane Society's Web site, www.hsus.org, states that over 4.7 million people per year are bitten in the U.S. by dogs—and most of these are children under 13 years. Be sure you and your children know how to behave around a dog, both those you know and those you don't. On this Web site are a couple of short videos on preventing dog bites. There are many reasons why dogs bite: out of fear; rough play or handling; protecting their territory, toys, food, or puppies; being woken up from sleeping; and establishing their dominance. A dog might chase a child the dog sees as prey or try to herd the child by nipping.

Teach children to respect their dog—no tail or ear pulling, no wrestling or climbing or jumping all over, don't poke it, annoy it, or back it into a corner. A baby brother would fight back with a sibling who did this by hitting or punching . . . a dog uses its teeth.

If your dog does bite, even if not a serious bite, do not ignore it, because it could lead to something worse. Call in a behaviorist to visit the dog at your home. She will help you assess the situation and help correct any bad behaviors. Be safe, be aware, all dogs have teeth, and sometimes they use them.

April 21 - The Vegan Voice

An ounce of action is worth a ton of theory.

—Friedrich Engels

Being a vegan I was thrilled to hear about a national crusade to urge ballparks, zoos, movie theaters, amusement parks, schools, hospitals, fairs, and other venues to serve veggie meat alternatives. Even people who are not yet fully vegetarian like to eat veggie dogs and burgers rather than the dead things.

The *www.soyhappy.org* approach encourages us all to speak out. Local consumer feedback is taken very seriously and if the venues think there's enough of a demand, they'll meet it.

Never think your one voice doesn't matter, it does when added to others to form a chorus. Please be gracious in your request—the nicer you are, the more they'll go out of their way to please you. Always ask if the concession manageress is available to speak to since she is the one who orders!

Let's all be soy happy and change the way our country eats . . . one happy venue at a time.

April 22 - Celebrate Earth Day Every Day!

Today is Earth Day, celebrated annually in over 184 countries in the effort for a healthy environment and a peaceful planet. Earth Day inspires grassroots communities to hold events and actions to educate and spread awareness. Earth Day celebrates our connection with nature, bringing an awareness that each one of us is responsible for the destruction or abundance of our natural world . . . the only one we have. Earth Day makes us realize that each of us has a voice and every one of our actions matters; collectively great things happen.

Celebrate Earth Day by joining an organized group and help clear beaches, parks, and wasteland of cans, paper, plastic, bottles, trash. Go out with your family and friends, or go out alone. Look around in your neighborhood to see what needs to be done, petition your local government for more trees, cleaner waterways, and an end to industrial pollution. Use earth-friendly chemicals; recycle paper, cans, and green stuff.

Log onto *www.earthday.org* or check your local press to find out what is happening in your area, how to organize your own event, or what commitment you could make in your own vital way to help save our Earth . . . and then make every day Earth Day.

April 23 - The Healing Power of Plants

In the rainforest areas the indigenous people still hang on to their old way of life and their effective, natural healing methods, handed down through their generations. These people still realize that the body, emotions, environment, brain, and spirit are all connected and must work in harmony to be healthy. They also still use the active ingredients in their plants to help them heal. How much we could learn from these people if we only knew how to listen.

Ethnobotany is the study of the relationship between plants and humans; scientists have discovered many drugs we use every day from the plants the botanists have brought back from jungle areas: quinine, morphine, and other plant chemicals to treat and cure. Scientists know how important rainforests are and that there are many, many cures to be discovered there. In Southeast Asia the traditional healers use 6,500 plants for healing and in Northwest Amazon the peoples use over 1,300 plant species for their medicines.

The rainforests are beautiful and necessary to the people and animals who live there, but we are causing their extinction for our own self-centered ends: cheap burgers and teak. Ninety-five percent of Brazil's Atlantic Rainforest has already been destroyed . . . seventy per cent of the plants there are found nowhere else on Earth and it's estimated that over 90 different Amazonian tribes disappeared in the twentieth century alone; all because of our greed.

Start caring. Make yourself aware about these primeval jungles, before it's too late.

April 24 - An Early Cure

Ginimony likewise burnt, and pulverized, to be mingled with the juice of Lymmons, sublimate Mercury, and two spoonefuls of the flowers of Brimstone, a most excellent receite to cure the flushing in the face.

—**Charles Kingsley**

April 25 - Your Skin Prefers Natural Makeup

What woman doesn't feel better after she's applied a little mascara and lipstick, a sweep of blusher and a dab of powder? When you look back in history, why even the men used to wear makeup to whiten their faces, enhance their cheeks and lips with color, and add a black dot for a beauty mole!

In Queen Elizabeth I's time—and we've all seen pictures of her bright-white face—pale skin was a sign of nobility and wealth, since only the poor peasants worked in the fields and were exposed to the sun. White foundation was applied to the neck and bosom and was achieved by a mixture of white lead and vinegar, but it left the skin "grey and shriveled." They also used mercury as a drying agent and ambergris (from sperm whales), civet (from polecats), and musk (from the male musk deer). One of the most appalling aspects of 16th century makeup was the poisonous nature of many of the cosmetics. But we know better today . . . don't we?

It seems not. Mainstream cosmetics are often derived from petrochemicals that can clog the pores, and they contain artificial colors and preservatives. They are not necessarily harmful but they can cause skin sensitivity, allergic reactions, and irritation. So, why not choose natural cosmetics that use herbs, plant oils, and nutrients to benefit your skin? Do some research of your own by checking your local health food store, throw out those nasty old products you've had for months, and choose natural makeup.

April 26 - Don't Compromise

Don't compromise your health or your skin by buying cheap, nasty cosmetics. Don't compromise an animal's health by choosing products tested on animals.

April 27 - The Many Uses of Mint

There are over 600 types of mint—a variety for every garden, every use, and every flavor. Mint is supposed to love fertile, well-drained, moist soil with a little sun. Try telling mint that—it decides it will flourish anywhere and everywhere! In a garden, mint is best grown in pots or it will gaily invade every flowerbed and you will be pulling up its roots for many years to come!

Mints have long been grown for their culinary and aromatic qualities. They are rich in essential oils, especially menthol with its antiseptic and refreshing qualities used for toothpastes, mouthwash, and teas and for its digestive qualities in antacids and stomach remedies.

My favorite mints are variegated pineapple mint, fuzzy apple mint, chocolate, orange, ginger, spearmint, and plain old peppermint. I like to add mint leaves to my salads, soups, baby potatoes, fruit punches, mint juleps, cakes, cookies, and scones. Or pick bunches of herbs and place in a pretty vase in your bathroom and kitchen. Be generous with mint, it will keep growing back.

April 28 - A Mid East Veggie Treat

Tabbouleh Boats

2	cups boiling water
1	cup uncooked bulgur
1	medium tomato, finely chopped
1	cup cucumber, peeled and diced

Dressing

2	tablespoons olive oil
1	cup fresh parsley, finely chopped
½	cup fresh mint, finely chopped
2-3	scallions, finely sliced
½	cup fresh lemon juice
	freshly ground black pepper, to taste
4	ripe avocados, halved, pits removed, to serve

In a medium saucepan, bring the water to a boil and stir in the bulgur. Remove from heat and set aside for 20 minutes. In a small bowl, whisk the dressing ingredients. Drain the bulgur and rinse, and then drain again. In a medium bowl, mix the bulgur with the tomato and cucumber and pour over the dressing. Toss well. Serve in the avocado halves as a nice light lunch for your friends. Serves 8.

April 29 - Planting Trees is Good For All

In 1872, J. Sterling Morton came up with the idea that every year a special day should be put aside for tree planting. Thus he founded Arbor Day and that year over a million trees were planted in Nebraska. Probably now, more than ever, we need to honor Arbor Day, which is usually celebrated on the last Friday in April (some states choose different dates depending on their weather for best tree planting times).

Trees hugely improve the quality of our lives. They provide shelter and food for our wildlife, clean the air, absorb carbon dioxide and release oxygen, they mask noise, prevent soil erosion, provide paper and wood for fuel and buildings, all this plus the joy and wonder of such a majestic and wonderful plant. Just imagine walking into a park or driving along the highway and there are no trees!

Celebrate Arbor Day by logging onto *www.arborday.org* and finding out what you can do in your area. Save a tree by recycling paper. Plant a suitable tree in your garden or neighborhood and dedicate it to someone special. Happy Arbor Day!

April 30 - The Herbal Home

Bring some cut, fresh herbs into your home. Rosemary, mint, lavender, and bay are all excellent as cut herbs, but be sure and change the water regularly or it gets a bit green.

Herbs always do well when cut back regularly whether as an indoor arrangement, to hang upside down to dry, or to use in your recipes. I always give them a good shake before bringing them in to dislodge any greenfly or other bugs.

If you have a baby you might want to cut some lavender, dill, and rue since, on April 30—May eve, or Beltane—mothers used to hang these herbs, along with garlic, near the cradle to keep the fairies from making off with their baby.

MAY

5

Myself in Nature

May 1 - Coming to Full Bloom

Today is May Day, which is still celebrated in many old British villages as a pagan day, with stilt walkers, a May Queen, Garland King, jousting 'obby 'osses and maypole dances. The young women and men of olde Britain would go out to the woods, chop down a suitable tree, remove the branches, and take it back to their village. There it was decorated with garlands and ribbons for all to dance 'round as a symbol of fertility.

It's always more fun when May Day falls on a sunny weekend, when the villages are even more idyllic with stone and thatch cottages, cream teas, elderflower wine and craft shops, hawthorn in full bloom, the children and their flower posies, and the strange Morris men in their costumes of white and red cavorting on the village green, the bells tied to their legs jingling as they dance around.

I am a big believer in keeping the traditions; they make up the fabric of our lives. If your town has a tradition at any time of the year, be sure and support it.

May 2 - Bathing in Tranquility

If you have a lavender field in your area, visit it on a blistering sunny day. Let the lavender's aroma invade your pores, let it bathe you in tranquility, and buy some to take home. Make a lavender wand or fill fabric bags to place in your lingerie drawer—surround yourself with lavender and evoke summer's spells.

May 3 - Lavender Relaxation

Lavender is a wonderful herb: relaxing, calming, and containing an essential oil that lingers evocatively.

May 4 - Ward Off Osteoporosis

Osteoporosis is a disease you often don't know you have until it's too late, and men can suffer from it too. We must all work to keep a strong framework for our bodies: here are some things to be aware of:

• Eating plenty of fruits and vegetables increases bone density. The more acidic the conditions in the body, the more calcium is lost from the bones. Fruits and vegetables help restore the body's alkalinity, thereby protecting the bones from calcium loss. Green vegetables, such as broccoli, spinach, cabbage, and kale, especially are good.

• Remember your vitamin D, much of which we obtain from sunlight on our skin—30 minutes a day on our bare arms and legs is good.

• Get enough calcium and magnesium, again from those green leafy veggies, nuts, seeds, beans, and other legumes.

• Eat soy foods. Osteoporosis is more common after menopause, a time when our bodies tend to lack estrogen, which has been proven to keep our bones healthy. Tofu, soymilk, soy cheese, and tempeh are all high in natural estrogenlike isoflavones.

• Exercise—lift weights, jog, walk, and do aerobics—just 20-30 minutes every day. It will help safeguard the strength of your bones.

People who may be prone to osteoporosis, such as people who smoke, have a tiny frame, are genetically inclined, etc., may also benefit from taking a bone-building supplement available from a health food store.

May 5 - Cinco De Mayo

Today we celebrate the 1862 victory of a small Mexican militia over the invading French army. Although more widely celebrated north of the border than south, Cinco de Mayo is a day for rejoicing in the triumph of indigenous peoples over invading forces.

May 6 - Celebrate

The Whole Enchilada

¼	cup soy sauce or tamari
2	tablespoons peanut butter
2	tablespoons tomato paste
1	tablespoon olive oil
1	package (12 ounces) firm tofu, drained and cut into ¼ inch cubes
1	onion, chopped
1	green bell pepper, chopped
1	green chili pepper, seeded and chopped
1	cup defrosted frozen corn
12	small flour tortillas
2	cans (15 ounces each) chopped tomatoes
1	tablespoon fresh oregano
1	teaspoon dried oregano

Be the Change You Want to See in the World

1   can (15 ounces) black or red beans, drained and rinsed

2   cups grated vegan cheese

Heat oven to 350°F. Lightly oil a large baking dish. In a large bowl, stir the soy sauce, peanut butter, and tomato paste until blended. Heat the oil in a large frying pan and stir in the tofu, onion, green pepper, and chili pepper. Sauté until softened. Stir this into the soy sauce mixture. Stir in the corn. Holding the tortillas with tongs, heat by passing over a gas flame until warmed and flexible. (Or heat in a microwave.) Place ½ cup of mixture in the center of each tortilla, roll up tightly, and place in the baking dish, seam side down. Continue with all the tortillas. Using the same bowl, stir the canned tomatoes, herbs, and beans together, then tip over the enchiladas. Sprinkle with vegan cheese. Cover the dish with foil and bake until bubbling for 30-35 minutes. Serve hot with a fresh salad and a bowl of salsa. Serves 4-6.

May 7 - Animal-Friendly Gardening

So many gardeners have companion animals sniffing around their flowerbeds, chewing on blades of grass, running and rolling across the lawn. But we need to take care that no harm comes to them. Be sure the fence is high enough that your dog can't get over it and secure enough that he can't tunnel under it, especially if he's the digging kind! Be sure latches are secure on gates and that children firmly close the gate behind them. If you have a pool or spa, be sure Rover is trained to avoid it or has access to climb out if need be. If he's old, never leave him unsupervised in the pool area . . . he could drown.

Set aside a little area in the garden for your dog, where he can bury his bones, dig a cool hole to lie in, and where he is allowed to go to the bathroom, all without you getting upset.

You must also be careful that the plants you choose for your flowerbeds are not poisonous if he chews them—Rover doesn't know they are toxic. Log onto *www.aspca.org* for a list of poisonous plants you should be aware of. By the same token don't leave sharp gardening tools lying around, or any fertilizers or chemicals.

Be safe and sensible with Rover and you can both enjoy your garden.

May 8 - Making Friends with Snails

If snails ever bother the plants in your garden, ask them to stop. Snails are quite responsive to kind requests, or, if you have stubborn snails, make a compost pile just for them.

May 9 - Take Time for Love

If you are married or coupled, make sure to show your love and care for your beloved today. And never take that relationship for granted. Every single day, do something to make that person feel special and loved.

Pretend you are on a game show and you know all the answers about him: favorite meal, after-shave, actress, movie, sport's star . . . be interested. Make an effort to be "one of the guys" occasionally—go for a drink with him, watch a ball game. Support him, listen to his opinions, snuggle on the sofa, be interested in the car, and help him wash it. Find activities you can do together. Wear pretty underwear and a perfume he likes.

Make an effort, be kind, and be loving and supportive . . . you'll have a happier, nicer relationship.

May 10 - Love in the Form of Flowers

How about buying some flowers for a woman in your life?

With Mother's Day near, the flowers in the shops are at their best. Try a bunch of all one type. blue irises, all peach or red roses, a highly perfumed spray of star-gazer lilies (be careful with star-gazers, women either love their smell or hate it!), bright carnations, tiny primroses, or fragrant freesias.

If they're wrapped in nasty cellophane, rewrap them in pretty tissue or brown paper with a bright ribbon or raffia. Hand them over personally if you can, to have the pleasure of the receiver's smile and a hug and kiss in return.

May 11 - Extending the Life of Cut Flowers

When you *receive* beautiful flowers, do you know how to best prolong their lovely lives? Here are some tips:

• Recut the flower stems on the diagonal with a sharp knife or scissors to expose a larger surface from which the flower can drink. Place them into water immediately.

• Remove any leaves that would be under the water to keep it fresher. When the water gets cloudy that means there are microorganisms present so change it every three days or so, and use the floral preservative that came with the flowers.

• Put the vase of flowers in a cool place, not in direct sunlight or on top of the TV, but still where they can be viewed and enjoyed.

May 12 - Mother's Day

There are all kinds of Mother's Day feasts. The Rice Krispies treat your children bring you on a tray, pinning you to the bed until you've finished every bite, is wonderful in its own way. But why not follow it with a truly spectacular brunch?

—**Mollie Katzen**, artist and cookbook author

Mother's Day falls on the second Sunday in May—right around now. Everyone takes mom out for brunch; why not show your love this year by treating your mom to lunch at your place? If you have a garden, set up tables outside and get the whole family to help.

May 13 - A Delicious Recipe

Here is a recipe to give you an idea for mom:

Artichoke Bruschetta

1	red onion, finely chopped
1	red bell pepper, seeded and finely chopped
1	clove garlic, crushed
1	tablespoon olive oil
1	jar (14 ounces) artichoke hearts, drained and finely chopped
½	cup parsley, chopped
½	cup pimento-stuffed green olives, finely chopped
2	teaspoons lemon juice

Be the Change You Want to See in the World

freshly ground black pepper, to taste

baguette, cut diagonally into 36 pieces, ¼ inch thick

½   cup vegan Parmesan

In a medium pan, sauté the onion, bell pepper, and garlic in the oil for about 10 minutes. Stir in the artichokes, parsley, olives, and lemon juice and season well. (Can be made a day ahead to this stage.) Arrange the baguette slices on a baking sheet in a single layer and broil on one side only until toasted. Turn the bread over. Spoon the artichoke mixture onto the untoasted side. Sprinkle with Parmesan. Place back under the broiler for about a minute until the cheese bubbles and the mixture heats through. Serve immediately. Makes 3 dozen appetizers.

May 14 - Woman's Comfort Food

Creamy Leek and Celery Bake

4   tablespoons vegan margarine

1   leek, rinsed and chopped

4   stalks celery, sliced

8   medium white mushrooms, sliced

      freshly ground black pepper, to taste

2   tablespoons whole wheat flour

2   cups plain soymilk

      homemade whole wheat breadcrumbs

¾   cup chopped pecans

1   tablespoon fresh parsley, chopped

Preheat oven to 350°F. Lightly coat an ovenproof dish with margarine. In a medium pan, melt 1 tablespoon margarine and sauté the leek, celery, and mushrooms for 5-10 minutes until softened. Season with black pepper. Stir in the remaining 2 tablespoons margarine and the flour and cook for a couple of minutes. Then pour in the soymilk and stir constantly to prevent lumps until the sauce is thickened and creamy. Tip into the oven dish. Using the same pan (to save washing up!) mix together the breadcrumbs, pecans, and parsley and sprinkle over the veggies. Bake for 15 minutes or until bubbling. Serves 2-4.

May 15 - Fresh Fruit = Strong Bones

Tropical Fruit Salad

2	bananas, peeled and sliced
1	orange, peeled and segmented
	small can (6 ounces) pineapple pieces, in natural juice
1	kiwi fruit, peeled and sliced
1	carton strawberries, sliced (or use raspberries)
4	ounces ginger ale

Put all the fruits in a pretty serving bowl. Mix gently but well. Pour over the ginger ale and serve immediately. Serves 4-6.

May 16 - Create a Butterfly Garden

Have you noticed there are fewer butterflies around now than from your childhood? Weed patches and natural meadows are disappearing to build shopping malls, and roadside stretches are sprayed to keep down the flowering weeds these vivid winged ballerinas live on.

Encourage butterflies into your garden by selecting the plants they like, not those that will impress the neighbors! You could set aside a part of the garden to look a little like a meadow. Persuade sulphur yellows, swallowtails, painted ladies, and monarchs to brighten your day. Here are some of the plants they love, but check your garden center to see what is suitable for your zone: The buddleia, or appropriately named "butterfly bush," would look great in a sunny corner with its lilaclike mauve flowers. Lantana comes in many colors, as does phlox. Black-eyed Susan (*Rudbeckia*) loves to ramble so combine with Mexican sunflower and purple coneflower for height. I also have lavender and pineapple sage, not only for the butterflies, but for the bees and hummingbirds too. Let your herbs go to flower for the sake of the butterflies and be sure you keep your garden organic. No chemicals now!

Don't forget butterflies like to drink: I put down a large saucer and half fill it with gravel and sand then check it each morning to keep it moist.

May 17 - Changing What We Can

In our current world, animal products are everywhere. It's impossible to live a 100% animal-free existence.

The film in my camera and on videos contains gelatin. Dyes that have been tested on animals are used in clothes I wear. And on and on it goes.

All we can do is our best and educate ourselves as much as we can on where cruelty exists and change what we can.

May 18 - Bearing It All

This week is set aside for bear awareness. To quote Defenders of Wildlife Web site:

> There are eight species of bears around the world, three of which—the polar, grizzly (or brown), and black—are found in North America. Bear Awareness Week will highlight these amazing creatures by helping to educate the public.

Defenders is working to help people and predators live in harmony. Log onto *www.defenders.org* for lots of bear facts, a scavenger hunt, and information about how you can "adopt" a bear. This is not just our country; it's their country too.

May 19 - Animal Rights

Even when a law has been passed our wildlife still isn't out of harm's way. If it weren't for organizations like Defenders of Wildlife I doubt almost anything would be safe. Alaska's Arctic Refuge, the manatee, lynx, wolves, otters, and now the Endangered Species Act all are under threat.

On the Defenders of Wildlife Web site is a link where you can join for free wildlife updates, called DenLines. Here you are informed of any shenanigans going on and of how you can help . . . simply by using your mouse to make yourself heard. You can "adopt" an animal or two (also makes a great present) or you can become a member and receive their excellent magazine. Defenders continues to need our support to carry on their fantastic work on our behalf . . . and on behalf of our wildlife and our wild places.

May 20 - Ending Cruelty for All Living Things

It is totally unconscionable to subject defenseless animals to mutilation and death, just so a company can be the first to market a new shade of nail polish or a new, improved laundry detergent . . . It's cruel, it's brutal, it's inhumane, and most people don't want it.

—**Abigail Van Buren**, "Dear Abby"

May 21 - Cosmetic Check

Over and over again, rabbits are being brought into the laboratories to face terrible agony under the false excuse that animal testing protects the health and safety of the American consumer. Manufacturing companies point out that product testing is necessary to ensure the safety of their products, yet it is not required by the Federal Food, Drug, and Cosmetic Act. We know that perfume poured into your eye may blind you— why keep doing it to a rabbit?

Put compassion on your shopping list. As PETA tells us:

> More than 300 companies, including the Body Shop, Paul Mitchell, and Aveda, manufacture safe, gentle, effective products that are tested, not on animals, but through in vitro (test tube) studies, with sophisticated computer models, and on human skin (cloned or still attached to volunteers!). Many companies are committed to using known-safe ingredients, (there

are now more than 600) rather than experiment with new chemical combinations.

Read the labels on any products you are about to buy. If they don't say, "Not Tested on Animals," keep searching until you find one that does. Write to the companies that test on animals and tell them just why you will no longer buy their products. With just a pen and paper, and the power of your wallet, you can be a strong advocate for animals. Boycotts are most effective *en masse*, so maybe circulate a petition in your workplace or school.

Get into the wonderful habit of choosing cruelty free.

May 22 - The Greenhouse Effect

Scientists already know what is causing global warming and we are all contributing to it: it is our wasteful attitude and short-sightedness. We burn too much fossil fuel and are contributing to massive deforestation of natural woodlands and forests. Fossil fuels are pretty much pure carbon, laid down by the Earth over thousands and thousands of years. According to Fred Krupp, president of Environmental Defense:

> Whenever you save energy—or use it more efficiently—you reduce the demand for gasoline, oil, coal, and natural gas. Less burning of these fossil fuels means lower emissions of carbon dioxide, the major contributor to global warming. Right now the U.S. releases about 50,000 pounds of carbon dioxide per person each year. If we can reduce energy use enough to lower greenhouse gas emissions by about 2% a year, in ten years we will "lose" about 10,000 pounds of carbon dioxide emissions per person.

Greenhouse gases are so called because when massive amounts collect in the atmosphere they trap heat, for 100 years or more, giving us extreme weather patterns by melting the ice at the North and South poles. This changes world weather, which further results in record droughts and record rainfall.

Here are some things you can do starting today:

- Support our scientists by letting our elected officials and congress know we need fossil fuel alternatives—wind power, solar power, and wave power.

- Choose more Earth-friendly transport, which also reduces smog-causing emissions.

- Recycle, conserve energy, and support the work of Environmental Defense and other environmental organizations.

- Go to www.environmentaldefense.org and become involved.

May 23 - New Diets for a New America

Erik Marcus's famous book *Vegan: The New Ethics of Eating* will help you understand why humans can no longer continue to eat meat and indulge in dairy from an ethical, ecological, and health viewpoint. Carol J. Adams's *Living Among Meat Eaters: The Vegetarian's Survival Handbook* contains polite rebuffs, suggestions for those who are dating or living with a meat eater, suggestions on raising veggie children, and how to survive summer barbecues, Thanksgiving dinner, and simple business lunches where your food choices may be under attack. Carol also includes fifty of her favorite recipes.

And then, of course, there's my bible: *Diet for a New America: How Your Food Choices Affect Your Health, Happiness, and the Future of Life on Earth* by John Robbins. I bought this years ago based on a rave review in *Vegetarian Times* magazine:

> When I finished reading Diet for a New America, I knew that in my hands lay one of the most profound studies ever written of how our eating habits affect our lives, and indeed all of life on our planet. If you read only one book this year, let it be this one.

May 24 - Every Little Thing Counts

Even if you don't go completely vegan, try to ease off animal products by adding some yummy vegan recipes to your repertoire. These days there are lots of vegan cookbooks to choose from. PETA's *Celebrity Cookbook* is chock full of recipes from famous vegetarians. There's Sir Paul McCartney's "All You Need is Avocado Soup," Jackie Chan's "Corn Chowder," and Alicia Silverstone's "Steamy, Creamy Artichoke Dip." There are recipes from Fabio, James Cromwell, Rue McClananhan, and Bridget Bardot, and it's all compiled by PETA founder, Ingrid Newkirk. My tummy was growling just browsing this book! But the thing I remember most about it was a quote from PETA director of Vegan Outreach, Bruce Friedrich: I can survive without greasy chicken wings—but the chicken can't.

May 25 - Eco-Souvenirs

If you're thinking about an organized eco-tour to see rainforests, the Andes, or the Himalayas, www.ecologyfund.com may help. Before you go, educate yourself on that region's specialty:

Be the Change You Want to See in the World

when we went to Crete we bought the local Malia bright blue pottery, in Maui some print fabric, Sydney was where I bought a toy koala, and Mexico their bright red and yellow throws.

Whatever you choose, buy something that won't just be stuck in a drawer on your return: surround yourself with your vacation reminders and make a feature of them. Choose items that were never part of an animal: no ivory, skins, horns, turtle shell, bird feathers, but still support the local industry and trade. Look particularly for items women have made, such as weaving, so that you support the artisan's whole family.

May 26 - Mind the Drip

We take water so for granted. We leave the faucet running when we brush our teeth, overwater our gardens, wash the car too often, take deep baths every evening rather than a quick shower, or don't mend that dripping tap. Become water wise.

May 27 - Water, Water Everywhere, But Not a Drop to Drink

Now that the flowers are in bloom and the sprinklers are running, time for a few facts about water:

- One cup of water costs five times as much in a Nairobi slum as in an American city.

- Three gallons of water provide the daily drinking, washing, and cooking water of one person in the developing world . . . yet in the U.S. it flushes one toilet.

- Three gallons of water weigh 25 pounds. Women in Africa and Asia carry, on average, twice this amount of water over 4 miles . . . each and every day.

- 470 million people live in regions of severe water shortage. It's estimated, if nothing is done, that by the year 2025 this will increase sixfold.

- Roughly 1/6 of our World's population does not have access to safe water.

- 2 1/2 billion people (roughly 2/5 of the world's population) do not have adequate sanitation and, according to the United Nations, 6,000 children each day die from unsafe water and sanitation: that is the equivalent of 20 jumbo jets filled with children crashing every day.

May 28 - Liquid Gold

Water is precious. Appreciate it by conserving it.

May 29 - Know Thy Body, Know Thyself

Your body is your body; it's the only one you have. It is your responsibility to take great care of it, not a doctor's. It's easy to feel overwhelmed by all the health information out there these days, some of it contradictory. Please allow me to suggest a book to you that will make clear all the things that go on in a woman's body in an unambiguous and concise way: *The McDougall Program for Women: What Every Woman Needs to Know to be Healthy for Life* by John A. McDougall, M.D., is the best book I've ever read on women's health. It also includes over 100 recipes by his wife, Mary.

Be the Change You Want to See in the World

Whenever I want to know something about my body I check in on Dr. McDougall: what I should do about my hot flashes, what exactly is cholesterol, is it best to get my omega-3 quota from flax seeds or flax oil, and just what is a healthy blood pressure? I don't just want to know *what*, I want to know *why* and Dr. M explains it all. He also has a clinic in the Napa Valley in California, other books, and a Web site for more information at *www. drmcdougall.com*.

May 30 - Barefoot and Loving It

One of the nicest things about summer is shedding our shoes and digging out sandals, if not going barefoot—but are your feet up for it? Here are some tips to stop any foot shockers when the weather turns warm:

• Keep your toenails trimmed by cutting straight across; shaping them like a fingernail can lead to ingrowing toenails.

• Blitz hard, dry skin with an invigorating pumice stone work out or special hard skin file. Walk on the sand for a natural workout.

• Use special foot scrubs and moisturizers containing peppermint to cool, refresh, and soothe, or lavender to smell good and keep feet healthy.

May 31 - Healthy Feet Lead to Happiness

More treats for your feet:

• Revitalize tired toes, feet, and ankles with a firm massage to relax aching muscles, stimulate and improve circulation.

- Offer to massage a friend's feet, or your partner's, in exchange for massaging yours, or treat yourself to a professional pedicure each month.

If your feet are happy, so are you!

JUNE

Love Is in the Little Things

June 1 - The Cat's Meow

June is the ASPCA's Adopt-A-Shelter-Cat month. Millions of cats roam our streets; some looking for a kind person to take them in and give them a new home and some are feral and want nothing to do with people. This huge population is especially acute at this time of year since shelters are over-flowing with all the kittens born during the springtime.

Unfortunately the message of spay/neuter still hasn't got through to some people and out the animal goes when they realize she is pregnant—out she goes, where, if she is lucky, she will arrive at a shelter. Shelter cats make excel-lent companions.

However, just as cats give people love and companionship, humans need to care for their cat and provide her with an appropriate loving home. Careful research and planning is the key to adopting the right cat for your family, since adopt-ing one should never be an impulse decision and indoor felines can live up to twenty years.

June 2 - Bee Happy

To keep butterflies and bees happy, be organic.

June 3 - Organic Herb Gardening

Thyme to plant a few herbs. Herbs can be used for many things, in sachets, oils, and lotions, the bath, medicinal, and culinary use and perfuming the home. And they are easy to grow.

However, herbs can be as prone to garden pests as any other plant so here are some tips to keep your herbs as free from

Be the Change You Want to See in the World

diseases and pests as possible. Remove dead leaves and flowers as often as possible to keep fungal diseases at bay. Patches of rust can appear when there is high humidity or poor air circulation: this could be a sign your herb pots are too crowded together so rearrange them for ample space and ventilation. Red spider mites thrive, on the other hand, in hot, dry conditions: remove any leaves with webs and spray with water for a few days. Whitefly in my garden loves my chives and dill and thrives around the bases of the plants. I find this especially in late spring when temperatures are lower. I move the pots to a sunnier spot, especially an outside windowsill where the air circulation is improved and spray with a soap solution. Black fly prefers my chocolate mint on the topmost, newer growth where it's more succulent . . . these I pinch off in the worst cases and dispose of and then out comes the insecticidal soap spray again. My Italian parsley sometimes attracts leaf miners: these are the larvae of flies, beetles, or moths and I just pick off the affected leaves. Slugs and snails I pick off each morning.

Keep up with the bugs and you'll have fresh herbs to savor all summer long.

June 4 - Open Your Heart

If you are ready for a cat to adopt you, check out *www.aspca.org* for information, then please stop by a shelter or rescue for a cat, or a kitten or two. Every cat deserves a happy nine lives.

June 5 - Joyous Noise

Nurture yourself and your children with all kinds of music.

June 6 - Natural Herbalist

Herbs aren't just for people! Animals like herbs, too.

June 7 - Healthy Animal Instincts

I wish I had a grassed area for Smiler to play on—dogs love nothing better that to roll on their backs in the grass. If I miss his doggie signals to take him out for a grass-munching walk, he'll chow down on the herbs in the garden. His favorite is lemon balm (*Melissa officinalis*) which amazed me at first as it's very lemon flavored and I didn't think he'd like the taste, but it's good for anxiety and the digestion so I bow to his natural instincts. He also likes parsley in his evening meal and likes to rub himself along the rosemary bushes, which add shine to his coat and help itchy spots. If he feels his breath smells, he'll chow a sprig of mint—chocolate mint is his mint of choice and pineapple sage if his meal was oily—like after pizza! Lavender is also good for healing and is a flea repellent . . . but he hates the smell!

Dogs seem to have a "nose" for the herbs that are right for them. Pay attention and learn.

June 8 - Cradle Your Cat

Cats live longer, healthier lives indoors. If you have a cat, don't put it outdoors to have fun—put more cat fun indoors. Set aside time to play with her and give her attention; cats can be total love sponges if you'll just give yours the encouragement. Provide scratching posts, different kinds of toys, a pot of kitty grass for her to chew, a shelf or two at different

Be the Change You Want to See in the World

windows where she can safely watch the birds (the birds will be safer too) and people going by outside; the ASPCA Web site can give you more ideas. And don't forget to tell your neighbors and friends their cat should be indoors: we don't want their furry friend brought in in a black plastic sack.

June 9 - Bridal Botany

June is a glorious month when nature puts on its brightest smile and romance blooms.

JUN

Whether you are planning a wedding or just want the house to smell glorious, cut some herbs and bring them indoors. Check out their meanings, too: rosemary is for fond remembrance; sage for virtue; oregano is for joy and happiness; and lavender symbolizes devotion. Ivy, a traditional herb, represents constancy and fidelity. Experiment with herbs. They bring the outdoors in and are filled with history and meaning, too.

June 10 - Nature's Pest Control

This is the time of year for outdoor parties. If your host is a gardener, take a container of ladybugs (*Hippodamia convergens*), which you can buy at garden centers, as your house gift, and tell her that one little bug will happily munch through around 400 of her greenfly in a week. Or choose praying mantis instead; hire an aphid killer and let nature rule the garden and not pesticides.

June 11 - Be Mindful of Your Pets

As you get ready for those days at the beach, for having friends round for vegan barbeques and soy ice cream, remember to take especially good care of your companion animals. Tomorrow I'll tell you more.

June 12 - No Hot Dogs, Please

Dogs can very easily suffer from heatstroke. Signs include excessive panting, whining, agitation, staring or glazed eyes, vomiting, and collapse. A dog that isn't treated for heatstroke may die.

Never leave your dog in your car, even with the windows cracked. If you see a dog in a parked car on a hot day and you think he may be suffering from heatstroke, call your nearest animal shelter or the sheriff's department.

Limit Rover's physical activity to the early morning and evening while it's cool. The sidewalk and hot sand will burn your best friend's paws.

During the day, make sure he has plenty of shade if he's outside—and that's all day, especially at midday. A doghouse does not provide adequate shade—it's a stuffy box—crawl in and see! Nor is the garage. Shade is a large tree with overhanging branches, a row of hedges, or constructed shade. If he's indoors, make sure his area is well ventilated and it doesn't catch the hot afternoon sun.

A large tub of water will stay cooler than the usual bucket; make sure it's spill proof and plentiful. Keep a hose outside and a child's wading pool for the dogs to slosh through; dogs

Be the Change You Want to See in the World

love it—some just walk through to cool their paws, some fish out the toys, some lie in it and try to do doggy-paddle! But all have a good long drink!

Some dogs like to play in the sprinklers, so if your dog is outside for most of the day, maybe set your sprinkler timer system to come on for a few minutes during the hottest part of the day—Rover can play and then lie down in the cool, wet grass or concrete.

One more thing I ask—please keep an eye on your neighborhood dogs. If you suspect a neighbor's dog is suffering from the heat or has no water, please politely tell your neighbor—they may not realize there is no shade at midday. If you need any help or advice on how to handle this, please call your local animal shelter where one of their staff should be happy to help.

June 13 - Make Your Choices for the Children

Did you know that, according to the Children's Defense Fund, in this affluent country there are currently 13 million children who are hungry or at risk of hunger? No, that is not a typing mistake—thirteen million—and for every four people who stand in line at a soup kitchen one of those will be a child.

Meanwhile billions are spent on wars by this country every year. Billions are spent on space research and space travel.

Families in this country and children in this country should not be experiencing hunger and poverty, not in the short-term, and certainly not in the long term. Let's reprioritize children over weapons and ending hunger over eating meat.

June 14 - Body Image Starts at Home

Our weight seems to increase as we get older, but we should be able to compensate by being wiser about what we eat. And besides, it isn't so much the fat we see that's the real problem—it's the fat we can't see that's trapped around the heart, other organs, and being carried in the blood that is more likely to cause a heart attack.

So forget the diet pills and fad diets. The best diet to lose weight is also the diet that will keep you healthy and that's a vegan diet: a diet based on highly nutritious foods of beans, grains, vegetables, and fruit and which are naturally low in fats and processed sugars, yet give enough protein for a healthy body. The key really is low fat since fats are twice as high in calories than are carbohydrates and proteins. This is a diet of taste and balanced nutrition and does away with calorie counting . . . actually it's not a diet, it's a healthy way of life.

June 15 - Summer Health Food

Meal in a Bowl

4	cups fresh spinach, torn
1	cup cauliflower, separated into small florets
1	cup fresh mushrooms, sliced
1	cup cherry tomatoes, halved
1	small cucumber, sliced
½	small red onion, thinly sliced
1	can (16 ounces) garbanzo beans, drained and rinsed

Be the Change You Want to See in the World

½	cup coarsely chopped walnuts
1	cup cooked brown rice
	low fat dressing of your own choice
1	teaspoon toasted sesame seeds

Place spinach in a salad bowl; top with cauliflower, mushrooms, tomatoes, cucumber, onion, beans, walnuts, and rice. Pour the dressing over the top and toss well. Sprinkle with sesame seeds and serve. Serves 2-6.

June 16 - Ward off Pests

Also with the warmer days come the crawling and buzzing beasties! Take care of yourself or you might get eaten alive!

June 17 - Gentle Care for Bee Stings and Bug Bites

If, like my husband, you attract the mosquitoes and bees, here are some tips so that you don't have to go into hiding until fall.

A bee sting is where the bee leaves his stinger, or lancet, in your skin where the venom continues to be pumped in. So remove the sting carefully and quickly by flicking it out with a fingernail or the edge of a credit card. Don't pinch or squeeze the area, as this will make it worse. Bee stings are acidic so neutralize the area with baking soda mixed with water.

Wasp stings tend not to leave a barb but temporary, local pain plus a swelling and red mark. To ease the pain use an ice pack, an anesthetic spray such as Wasp-ese, and an aspirin if necessary. You can also dab wasp stings with vinegar to help the inflammation. If a baby or toddler is stung by either a wasp or bee, there are multiple stings. If you are stung in the mouth, or know you are allergic to these stings, seek immediate medical attention. If you see a wasp or bee, leave it alone and walk away from it, don't try to swat at it. Always check anything you are about to eat for bees or wasps and wash your sweet sticky hands, and children's faces, afterwards to deter insects.

Mosquitoes tend to hunt at dusk and their itching bites can drive you nuts. If your skin tends to react badly apply something soothing like ice, witch hazel, or an antihistamine cream. Don't scratch at it and break the skin or it may become infected. To deter mosquitoes spray yourself with an insect repellent, avoid using anything perfumed on yourself, cover up with long sleeves and pants, paying special attention to wrists and ankles where your blood vessels are close to the surface. Burn special night repellents or citronella candles if outdoors and remember to close windows and doors at dusk. Avoid sitting near fresh water ponds or streams at dusk, and in your own garden clear up any pockets of standing water where mosquitoes may breed and hang out.

Make up a special summer first aid kit with antiseptic wipes and the items I've mentioned to take on picnics or to the beach. (Jellyfish stings can be removed with a dry towel or dry sand and bathe with vinegar to neutralize.) Add to your kit some homeopathic remedies such as arnica for bruises, calendula cream for sunburn and bee stings, lavender and tea tree oils for stings and minor cuts, plus rose or chamomile oil for overheated skin and to calm excited children at the end of

Be the Change You Want to See in the World

a hectic day. (Don't use these oils directly on the skin.) Keep an extra cooler of water for any emergencies too.

June 18 - Share the Planet

It's thought that HIV that manifests itself as AIDS began in Africa as part of the bush meat trade whereby humans ate chimpanzees or gorilla meat and the virus broke out by switching to another species. People in Asia contracted the SARS (Severe Acute Respiratory Syndrome) virus maybe from eating civet cats, a delicacy there, and then global air travel and crowded urban living spaces helped spread the disease. We have all heard of Mad Cow Disease or BSE (Bovine Spongiform Encephalopathy, which is a chronic, degenerative disorder affecting the central nervous system of cattle) where overcrowded animal farming and feeding herbivores with a food containing ground up animal remains spreads disease, both to the animals and then to humans.

West Nile virus, hantavirus pulmonary syndrome, and Ebola—are these and other diseases borne from exposing new pathogens on our quest to delve deeper into the rainforests and jungle areas? Do these pathogens then flourish thanks to global warming?

In our greed for resources and quest for ever-expanding markets, we have thrown open the lid to Pandora's box and we really need to close it before we do irrevocable damage—to ourselves, to the animals and plants we share our planet with, and to our Earth itself. Are these new viruses Mother Nature's way of containing humans before we do even more damage?

June 19 - Even Vegans Crave Chocolate

Truffles, praline, chunky or wafer-thin, soft-centered, ganache, brownies, cake, or cookies—chocoholics love chocolate. It's even good for you, in moderation. Chocolate contains antioxidants, which protect cells and repair cell damage, even to the extent of lowering cholesterol levels and reducing the risk of cancer. It's also claimed that chocolate improves your mood, decreases fatigue and increases alertness, makes you calm, yet euphoric . . . let's face it, we all know that eating chocolate simply makes you feel better.

Choose good quality, organic vegan chocolate, if you can find it.

June 20 - Yummy on a Stick

Barbecued Tropical Kebabs
Kebabs

4	kebab skewers
1	mango, peeled and chunked
1	papaya, peeled, halved, seeded, and chunked
2	kiwi fruit, peeled and chunked
1/2	small pineapple, peeled and chunked
1	large banana, peeled and sliced
2	tablespoons lemon juice
2/3	cup white rum

Dip

4	ounces vegan chocolate chips
2	tablespoons light corn syrup

Be the Change You Want to See in the World

1   tablespoons unsweetened cocoa

1   tablespoon cornstarch

3/4   cup soymilk

Dip the banana slices into the lemon juice to prevent discoloring.
Thread the fruit pieces onto 4 skewers, place in a shallow dish, and
pour the rum over them. Leave to get drunk for 30 minutes—the
fruit, not you—until ready to barbecue. Cook over the hot coals until
seared—about 2 minutes, turning frequently. Serve with the chocolate dip: Put all the dip ingredients into a small pan and heat over the
barbecue or a low heat until smooth and thickened, stirring constantly. Serves 4.

June 21 - Decadent Frosting , Animal Free

Chocolate Frosting for your Favorite Cakes

1   packet (12 ounces) of Mori-Nu silken style tofu, extra firm

3/4   cup maple syrup

1   tablespoon maple or vanilla flavor

½   cup cocoa powder

2/3   cup almond or cashew butter

Complicated instructions here, so read carefully! In a food processor,
whiz together all the ingredients until smooth and creamy. Can be
stored in the refrigerator. Makes about 3 cups . . . and remember, if
you have chocolate on your hands, you are eating far too slowly!

June 22 - Fruit for Thought

Don't only think of strawberries as a dessert either: I had a delightful salad the other day of mixed baby salad leaves, strawberry quarters, medjool dates, and a wonderfully light strawberry dressing.

June 23 - Treat Your Body Right

Sweet Summer Salad

1	bag Romaine lettuce
¼	head green cabbage, thinly chopped
2	carrots, grated
2	celery stalks, chopped
10	or so strawberries, quartered
2	fresh nectarines, diced
2	fresh peaches, diced
1	apple, diced
1	pear, diced
½	cup fresh orange juice
1	cup slivered almonds or walnuts

Toss the salad leaves, veggies, and fruit in a large salad bowl. Pour the orange juice over the salad. Sprinkle with nuts and enjoy. Serves 4-6.

June 24 - De-Stress with a Furry Friend

Animals seem to protect people against the stressful events that occur in their lives. Having a companion animal improves the life of the elderly. Did you know that scientific research suggests that companion animals help reduce blood pressure, promote healing after a heart attack, reduce anxiety, and boost the spirits?

June 25 - Beach OutReach

The Surfrider Foundation is an international nonprofit organization dedicated to the protection and enjoyment of the world's oceans and beaches through conservation, activism, research, and education.

Local chapters of this nineteen-year-old group work hard testing our oceans' water, cleaning beaches, campaigning, and visiting our schools for the benefit of all people, and hence the marine life, and we must all be mindful of the part we play when we visit our coastal areas.

The Surfrider Foundation distributes an excellent leaflet, "20 Ways to Cleaner Oceans and Beaches"; call 1(800) 743 SURF, or check www.surfrider.org for a copy. The leaflet starts with storm drains that lead directly to the ocean, "Dumping one quart of motor oil down a storm drain contaminates 250,000 gallons of water," and ends with the respect of the beach and your whole environment.

June 26 - The Most Lucious of Vitamins
Peaches and nectarines contain calcium, vitamins A and C and a little iron. Buy the freshest you can find, and cut out any bruising.

June 27 - You're a Peach!

Peach Salad

2	peaches, pits removed and diced
½	yellow or orange pepper, diced
2	cups white mushrooms, sliced
1	small head Napa cabbage, sliced
½	cup sunflower seeds
½	cup slivered almonds
½	cup pine nuts
2	packages Ramen noodles, broken, uncooked (don't use the flavor packet)

Dressing

¼	cup sugar
¼	cup apple cider vinegar
¼	cup rice wine vinegar
¾	cup oil
	freshly ground black pepper, to taste

Five minutes before serving, place all ingredients in a large bowl, add the dressing, and toss gently but well. Check the seasoning. Serves 6-8.

June 28 - Be Kind to Your Fellow Man

In these days of rush, rush, rush, self, self, self, let's not forget good, old-fashioned common courtesy, especially when in our cars. Never run a red light because you are in a rush. And remember, the passing lane is for temporary use, it is not yours to remain in until you care to move over or you've finished your phone call. And that's another thing . . . mobile phones. Please speak softly and move out of the way when you are so engaged. No one cares how important you think you are; be more considerate.

June 29 - Center Yourself with Yoga

There are many reasons to take a yoga class: in response to aging where your muscles are seizing-up, to release stress and tension and learn how to relax in both body and mind, to build muscle and tone, to become flexible and strong, to learn how to breathe deeply, for calmness, focus, and centering, or for self-awareness and peace with yourself and your world. Yoga is much, much more than just stretching, it is a discipline and a balance. Originating in India, yoga is as ancient as it is modern and encompasses all religions, all philosophies and can be practiced by any age group and level of fitness.

There are different styles or traditions of yoga from Hatha, Viniyoga, Ashtanga, Kundalini, Raja, and others so you need to try various ones and different instructors to find one that is just right for you. Of course you may just want to go to a class because your friend does and it's convenient to where you live or work, and that's fine too.

June 30 - A Picnic in the Park

What is your idea of a great picnic? Of course you want the weather to be nice—not too hot, yet not windy either, no bees, wasps, or ants, peaceful if you want romance or to relax, or other families around if your kids love to play. Of course good food goes without saying so give a little thought to your preparation, make a list and keep it handy or in your picnic basket or cooler and your day will be perfect . . . maybe!

I like to use a wicker basket complete with reusable plastic plates, utensils, cups, napkins, hand wipes, and a blanket. Cold drinks and food go in my cooler.

The food you take should be easy to prepare and transport, easy to eat (nothing that'll go soggy, drippy, or limp please), and tasty to eat with your fingers or a fork. And don't forget the pepper and salt. Wash fruit and chop veggies before you leave. If you take wine remember the corkscrew! Rinse a washcloth and place in a baggie for sticky faces and fingers! Put a pinch of patience, a dash of humor, and a bucket of fun in a bigger baggy and above all, relax and enjoy the day.

JULY

Summer's Inspiration

July 1 - Relax, Rome Wasn't Built in a Day

This seventh month was named after Julius Caesar, who was born during this month. This is a month of bright mornings, delightfully long afternoons, and long, light evenings. Throw your watch in a drawer and change to beach and picnic time. Think Mediterranean for simple summer food—salads, olives, ripe tomatoes, fresh crusty breads, perfect peaches, strawberries, raspberries, and icy lemonade. Sit by the pool, invite your friends. Read poetry, giggle at nothing, create memories and wrap in a pretty shawl as dusk falls among the night-scented jasmine and nicotiana. Relax and enjoy your summer.

July 2 - Aviation Inspiration

Today marks the anniversary of the last day Amelia Earhart was seen alive.

For most of her life Amelia Earhart defied conventional behavior suitable for a woman. She was the first woman to fly the Atlantic solo, the first woman to fly it twice and cross it in the shortest time. (Lindbergh had last crossed the Atlantic solo five years earlier.) She returned home to a ticker-tape parade, honors of every kind, and was voted Outstanding Woman of the Year.

On June 1, 1937, just a few months short of her fortieth birthday, this strong, courageous woman set off from Oakland, California with her navigator, Fred Noonan, to begin her around-the-world flight, following a route as close as possible to the equator. They touched down first in Puerto Rico, then down to South America, across the Atlantic to Africa, on to the Red Sea, India, down to Singapore, across to northern Australia and then to Lae, on the island of New Guinea—22,000 miles later; they had 7,000 miles remaining. It was now June 29 and Amelia was ill and very, very tired. She refused to give up.

On July 2 at exactly 00:00 hours Greenwich Mean Time, their plane the Electra, left Lae bound for Howland Island in the South Pacific. Amelia and Fred were never seen again. This year let's celebrate July 2 as Amelia Earhart Day. Who knows? Maybe it'll catch on!

July 3 - Consciousness and Fireworks

Fireworks may be noisy fun for humans, but they petrify our companion animals. We know that fireworks light up the sky in a multitude of colors and then kaboom! But our animals have no idea what it is and those high pitched screams as the fireworks rocket into the night sky startles their sensitive ears, resulting in that fear that makes the animal engage its fight or flight instinct. Their adrenaline flows . . . right down to their paws and they either hide, or take off running—sometimes for miles. Try to keep your pets away from fireworks, please!

July 4 - Remember Your Furry Friends

Bring animals indoors and shut doors and windows. Even if they haven't been afraid before, be on your guard. That six-foot fence is nothing when adrenaline kicks in. Close draperies and blinds to form an extra barrier of safety—we had a Lab at our shelter because he literally ran at a window, smashing it and escaping. Put the TV on louder or turn on a radio to cover up the firework sounds. Remember to close the doggy door if your pet usually uses one.

Even though they are indoors, keep the ID tags on, and be sure Fido and Fluffy have been microchipped in case the collar comes off.

Don't take Fido to a fireworks' display—even scarier—and don't leave him in the car either—hot and scary.

July 5 - Respect All Life

Even little flying creatures are living things worthy of respect.

I was doing my hour-in-the-garden and found a smallish July bug floating in the spa. I quickly found a large dead leaf and scooped him from the water. He didn't move. Ever hopeful, I placed him on the wall in the event he may still be alive. An hour or so later, two more July bugs buzzed me, reminding me of the spa incident. I went to see him, but he hadn't moved. I turned, saddened, and saw Smiler staring into the spa—there was another, drowning in the water. I leafed him out and he started buzzing. Hoorah! I saved a life today. I placed him gently in the potted lavender and left him to dry out.

July 6 - There is Wonder in a Blade of Grass

Small, seemingly insignificant lives are precious in nature: it's why I'm vegan. Don't kill them. If you must, just move them out of your way or escort them outside. But remember, they are living things, too.

July 7 - Buy What's in Season

If there is a local farm or farmers' market near you, buy your produce there. Why? It should be fresher, maybe organic, it's what's in season, you are supporting your local economy and helping the environment by lessening trucks driving across the country, or shipped from abroad. The produce should also be less expensive—no middle man, no cashier, no credit cards, and you take your own bags. If you go first thing, as soon as they have set up, you'll get the best choice. If you arrive last thing, you should get some good bargains, as they certainly don't want to repack and cart it all home again.

July 8 - Boycott Puppy Mills

Did you know that nine out of ten puppies sold at pet shops come from puppy mills? What is a puppy mill? They are commercial breeding facilities that mass-produce puppies at the lowest possible cost, for sale in pet stores. There are thousands of puppy mills in the U.S. and, with a few exceptions, the dogs are kept in overcrowded and filthy conditions.

Please don't support this inhumane practice.

July 9 - Avoid Pet Store Purchases

So, what can you and I do to stamp out these awful puppy mills?

Ask everyone you know not to buy a puppy from a pet store. Without pet shop sales, dog breeding would not be a lucrative business and most mills would be forced to close.

If you buy your pet supplies from a shop that *does* sell puppies and kittens, please consider changing to a store that doesn't and inform the original store why you have changed. Ask them to stop selling puppies (and kittens—yes, there are "catteries" too). Fortunately, some pet stores refuse to patronize puppy mills and catteries and have chosen to serve as off-site adoption facilities for local animal rescue groups.

If you have your heart set on a puppy, do some homework. Contact breed rescue groups. To find rescue puppies or a reputable breeder who breeds for good health and temperament, where you can visit the puppies at home, you can check out the mom and maybe the dad, check their health and check the puppies are well taken care of, socialized, and loved. And don't forget to phone round all the shelters in the area too—40 percent of all dogs in shelters are pure bred. Check Web sites for reputable adoption agencies—there are thousands of wonderful animals just waiting for your phone call.

July 10 - Look Up for Inspiration

In ancient times, before we messed up our atmosphere with smog, our night sky was unobscured and our ancestors drew images in the night sky by connecting the dots of stars. These are now called constellations, but to them they were star pictures: pictures of bears—Ursa Minor and Major—a bull or Taurus, a plough, twins (Gemini), and a big dog called Canis Major. The brightest star of Canis Major is called Sirius and "dog days" occur when the Dog Star, Sirius, passes closest to the rising sun at dawn and continues for about 40 days. These "dog days" fall between July 3 and August 11—and the heat at this time is due not to the added radiation by Sirius to the sun as the ancients thought, but as a direct result of our earth's rotational tilt.

Enjoy the dog days and do a little stargazing of your own. It's good for the soul.

July 11 - Take Skin Cancer Seriously

With summer full upon us you're probably out in the sun quite a bit, but are you dutiful about applying that sunscreen? Skin cancer is serious business and frighteningly common. There are three main types of skin cancer:

1. *Basal cell carcinoma*: The most common form and the most common of all cancers, it affects about 800,000 Americans every year. Chronic exposure to the sun causes almost all basal cell carcinomas (BCC's) and you'll usually find them on those areas most exposed to the sun—face, arms, neck, shoulders, ears, scalp, but they can turn up on nonexposed areas too. People with fair skin have a higher risk, as do those who work outdoors, or who spend a lot of their leisure time in the sun, such as surfers. Warning signs vary, so when you examine your skin and see a reddish patch—whether it itches or not, an open sore that won't heal, a shiny bump, a new mole, or anything suspicious—make an appointment to see a skin specialist . . . yesterday. Log onto *www.skincancer.org* for photographs and more information.

2. *Squamous cell carcinoma*: This affects over 200,000 Americans a year. They are found both on areas exposed to the sun—the face, ears, bald scalp, arms, etc., but also on other areas of the

body, including inside the mouth, lips, and nose. They begin in the upper layers of the skin (the epidermis), but can spread (metastasize) to distant tissues and even organs.

3. *Melanomas and moles*: Nearly everyone has moles, freckles, or birthmarks—brown marks on their skin—some are normal but some may be cancerous. Some are on daily plain view; some are hidden by clothing or by hair. Be aware of yours, know your body, and know what to look for. If a melanoma is caught at an early stage and treated, it is most times curable. If a mole on your body gets larger, changes its coloration, or the skin around it does, it begins to itch or feel tender (but it may not, I'm afraid), or the surface elevates or becomes scaly, crusty, oozing, or even bleeding . . . see a specialist. Do not put it off.

Melanomas are the deadliest of the skin cancers and can metastasize to other parts of your body, but in its early stages melanomas can be readily treated. The cure rate keeps rising and less and less tissue is removed.

Along with your monthly breast examination include a skin self-examination and get an annual skin exam by a skin care specialist, too. Take skin cancer very seriously and go to *www.skincancer.org* for more information.

July 12 - Sunproof Your Skin

A few more facts about skin and cancer:

• The more that children are exposed to the sun, the greater is their risk of developing permanent skin damage and skin cancer.

Be the Change You Want to See in the World

- Sunproof your baby and your toddlers. Teach them good sun protection when they are very small. Show them you are doing it too—set an example.

- One sunburn with blisters on a child can double the risk of getting melanoma later in life.

- By preventing your child from getting sunburnt, you lower their risk of cancer.

- More teenagers and young adults have been diagnosed with skin cancer these past few years than ever before.

- You are never too young, or too old, to be harmed by the effects of the sun. The sun will also age your skin before its time.

July 13 - Oil and Vinegar for Summer Delight

At the height of summer, when herbs are at their flavorful best, I like to capture those oils to flavor oils and vinegars.

Making flavored oils and vinegars is really easy. You'll need attractive bottles—glass or pottery—that are free of chips or cracks with tight fitting lids or corks. (Pier 1 sells them in different sizes.) Sterilize them in a bath of boiling water for 10 15 minutes before using to kill harmful bacteria.

Use good quality oils and vinegars—I like white wine vinegar and champagne vinegar. Olive oils should be virgin or extra virgin. Decide next what you'll use the oil or vinegar for to determine the herbs you'll use. (See the recipes below for some ideas.) Place the desired flavorings into the sterilized bottles, tightly cap, and store to allow the flavors to infuse: flavored oils for a couple of days, vinegars store for three to four weeks in a cool dark place. Use within three months. Oils should be refrigerated and used within ten days.

July 14 - Olive Oil for Health

• Use olive oil as a dip for crusty, home baked, or Italian, breads.

• Drizzle herbed oils over soups and veggie stews to "complete" the dish and add an extra layer of flavor.

• Marinate veggies for the grill in flavored oils.

• Toss some al dente pasta in basil or oregano oil for a tasty, easy midweek meal.

• Drizzle oils over pizzas instead of cheese, or use roasted garlic oil in your mashed potatoes!

July 15 - An Italian Tradition

Italian Dipping Oil

	sterilized bottle or jar
3-4	cloves garlic
20	peppercorns (red, black, pink, or green)
1	sprig each basil, thyme, and oregano, well washed and patted dry
	enough virgin olive oil to fill bottle or jar

Place the garlic, peppercorns, and herbs in the sterilized bottle and fill with virgin olive oil. Refrigerate for 2-3 days. Use for anything Italian or Greek.

Be the Change You Want to See in the World

July 16 - It is Lemon Thyme Time!

Lemon Thyme Vinegar

>   sterilized bottle or jar (I usually use 12- or 16-ounce
>   bottles for this)
>
> 1   lemon
> 6   (3 inch) sprigs fresh thyme, well washed
>   enough white wine vinegar to fill bottle or jar

With a vegetable peeler, pare a thin spiral strip of peel, 6-8 inches long, from the lemon. Push this into your sterilized bottle with a chopstick or wooden skewer, along with the thyme (I have lemon thyme in my herb patch so I like to use that). Fill the bottle with white wine vinegar and seal. Store in a dark, cool place.

July 17 - International Basket Delight

Oriental Basket: No flavored oils here, just a bottle each of sesame oil and rice wine vinegar, plus chopsticks, a decorative fan, a bottle of sake and some little sake china cups, some noodles, soy sauce, herbal tea, and fortune cookies.

July 18 - Don't Squander Our Gifts

Water scarcity is a huge problem these days, yet still we pollute it, squander it, and totally take it for granted.

In many regions in the United States, saving water is mandatory—especially in the summer. To make every drop count, please consider every drop. Remember not to leave taps running unnecessarily, don't over-water the garden, and don't run loads of laundry or dishes before you have a full load. Yes, saving water is a hassle—it's not difficult to do, it's just that most people can't be bothered—but we must think long term. Don't waste a drop!

July 19 - Think Globally

For more information on what you can do to save water, check out *www.earth911.org*. They'll give you lots of facts on water wasters to watch. Let's all do our bit to save this very precious, life-giving resource.

July 20 - A Crop to Save the Planet

Hemp is an annual herb with impeccable environmental credentials. It can be grown almost anywhere, from Africa to Norway, and its many uses make it a viable cash crop for small farmers throughout the world. In fact, hemp grows better in organic farming as it smothers weeds and controls pests, improves the structure of the soil with its strong root system to help erosion, and returns nutrients to the soil as it grows. Unfortunately, hemp's reputation is clouded from its association with marijuana, though "industrial" hemp plants have negligible narcotic content.

This summer when out clothes shopping, if you see anything made from hemp, buy it! This textile is strong, durable, and very hard wearing, being three times stronger than cotton. Hemp and cotton T-shirts keep you cool in the summer, and the fibers block UV rays. Oh, and if you're out sailing, don't worry, hemp doesn't rot in salt water!

July 21 - Bird-Friendly Pest Control

The three most common pesticides in the garden—diazinon, chlorpyrifos, and brodifacoum—kill thousands of birds every year. Every time you spray your lawn or your bushes with pesticides, you are helping to poison wildlife, birds, and next-door's cat. You can protect your garden from unwanted insect pests with strong-smelling roots and spices such as garlic, onions, horseradish, ginger, cayenne, and other hot chili peppers. Puree a combination of these, strain, and dilute with water in a spray bottle, then spray away!

By the way if you use a general insecticide you are also killing beneficial insects that pollinate our flowers, feed birds, or who even eat their weight in bad bugs. Check out your local garden centers for pots of live ladybugs. They cost around $7.00 for loads. Spray down the foliage with water before releasing your ladybugs (or they'll fly away in search of water), and they'll happily munch away on your white flies, aphids, and other pests. And their larvae will eat even more!

July 22 - A Fruitful Marriage: Fresh-Flavored Lemonades

Blueberry Lemonade

2	pints blueberries
1	cup sugar
1	cup fresh lemon juice, from approximately 7 lemons
1	quart springwater
	fresh peach slices for garnish

In a food processor, puree the blueberries, sugar, and lemon juice until smooth. Strain the mixture thought a fine sieve into a pitcher and stir in the springwater. Pour over ice cubes in a tall glass. Garnish with a peach slice. Makes about 1½ quarts.

July 23 - Add Zest To Your Life
Raspberry Lemonade

1	cup fresh raspberries
2/3	cup sugar
1	cup lemon juice
2	cups springwater

In a small bowl, mash the raspberries with sugar. Let stand for 10 minutes. Strain through a fine sieve into a pitcher; discard the seeds. Stir in the lemon juice and water. Taste and add more sugar if desired. Makes about 1 quart.

July 24 - A Month of Celebration

This month celebrates a whole lot of things—if anything floats your boat . . . celebrate it too! The most obvious, of course, is July 4, but this month also celebrates National Picnic Month, National Blueberry Month, United Nations World Population Day, International Massage Week, National Ice Cream Day, Tour De France cycle races, and Bastille Day. Of course, what's to stop you from combining them all on one day. Pack a picnic of French bread and wine, cycle to a picnic spot, where your partner gives you a massage, and on the way home eat blueberry ice cream. Happy all of July!

Be the Change You Want to See in the World

July 25 - Get the Sting Out! Naturally

Around this time of year we all get a few stings, a sunburn, a cut, or bruise—all from summer play. The next time you get a bruise or a sting, before you go running to the drugstore, try homeopathy.

July 26 - Homeopathic Healing

Homeopathy dates back to the time of the ancient Greeks—in fact, it was written about by Hippocrates, the father of medicine. But it wasn't until the 18th century that Dr. Samuel Hahnemann, a German physician, realized the principles behind it. During his experiments he found that many illnesses could be treated by using minute doses of a substance that caused similar symptoms when given to a healthy volunteer: mimicking the effects of the illness apparently adds to the patient's ability to combat the illness. In fact, the Greek word homoeopathy means "similar suffering."

Homoeopathic medicines are available for a wide range of conditions from cuts and bruises to stomach upsets, coughs, colds, cystitis, nausea, indigestion, eczema, premenstrual tension, and others. Consult a qualified homeopath for severe symptoms and you can educate yourself for any minor illnesses you may have.

July 27 - Honor the Sacred Coyote

The coyote (*Canis latrans*) is a member of the dog family and is a native in California. If you've never seen one, they look very much like a small German Shepherd dog with a long snout and a bushy, black-tipped tail. If you live near some hills or in a canyon, you'll often hear their high-pitched yapping at night.

Coyotes have adapted very well, since humans are building on the coyotes' territory and they are able to survive on whatever food they can find available—mice, rabbits, birds, carcasses of dead animals, garbage, pet food left outside, companion cats and small dogs. A black pit bull mix was brought into the shelter once as a stray, we called him Hero—he'd been hit by a car and had what the vet said were coyote teeth marks on his thigh.

Coyotes have become used to people—losing their natural fear. Some have become so bold and aggressive they have even attacked children. Coyotes are beautiful creatures and they were here first; it is we who are encroaching on them and we need to honor and respect that. There is enough room for everyone and we need to keep coyotes wild; not shoot them, not trap them, not poison them, but live in harmony with them.

July 28 - Keep Coyotes Wild

How can you help keep coyotes wild and make sure they retreat from us?

- Do not leave out food for your companion animals.

- Keep a tight lid on your trash cans and don't leave out plastic bags that contain garbage.

- If you know there are coyotes in your area, do not allow small children to play unattended, even in your yard.

Be the Change You Want to See in the World

- Do not leave companion animals outside. When out walking in the hills keep your dog on a long leash, or in sight at all times—especially at dawn and dusk.

- Clear brush and plants from around your house—not only for fire protection, but to discourage cover for mice, rats, and ground squirrels, which will attract coyotes.

- If you see a coyote in your neighborhood or coming down from the hills, frighten it away—wave your arms and yell, spray with a garden hose if in your yard, throw something at it—you want the coyote to remain wary and afraid of humans. Let your neighbors know there is a coyote in your area—put up notices.

- If you do encounter a particularly aggressive coyote, or if you are concerned in any way, contact your local animal shelter or Fish and Game office for further advice.

July 29 - Crisps and Crumbles

With these long, hot days, I don't want to spend hours in the kitchen, but when friends come around for dinner, I do like to finish the meal with a nice dessert and make use of summer fruits.

Well, here are the recipes I love to serve, especially with soy vanilla ice cream. Feel free to change the fruit depending upon what is in season.

July 30 - An Apple Crisp a Day Keeps the Doctor Away

Raspberry Apple Crisp

3	cups Granny Smith apples, peeled and cubed (about 1 pound)
2	cups fresh or frozen raspberries, or use blackberries
1/2	cup brown sugar
1	inch fresh ginger root, peeled and grated
	pat of vegan margarine

Topping

1/3	cup whole wheat flour
1	cup rolled oats
1/2	cup brown sugar
1/4	cup sunflower oil

Preheat the oven to 350°F. Combine the apples, raspberries, ½ cup brown sugar, and ginger in a medium bowl. Coat an 8x8 inch baking dish with margarine. Spoon the fruit mixture into the pan. Using the same bowl, combine the flour, oats, ½ cup brown sugar, and sunflower oil and stir with a fork until crumbly. Sprinkle this over the fruit and bake at 350°F for 40 minutes or until bubbling. Good served hot or cold. Serves 4.

July 31 - Southern Pacific Comfort

Aloha Crumble

2	mangoes, peeled and sliced, or use frozen
1	papaya, seeded and sliced
1	small fresh pineapple, peeled and cubed
2	teaspoons ground ginger
½	cup vegan margarine
½	cup light brown sugar
1½	cups whole wheat flour
1	cup shredded coconut, plus extra for garnish

Preheat the oven to 350°F. Place the mangoes, papaya, and pineapple in a pan with ½ teaspoon of the ginger, 2 tablespoons of the margarine, and ¼ cup of the sugar. Cook over a low heat for 10 minutes until softened. Coat a 2-quart baking dish with margarine and spoon in the fruit mixture. Mix the flour and remaining ginger together, then rub in the margarine and stir in the rest of the sugar and the coconut. Spoon over the fruit to cover completely. Place in oven and bake for about 40 minutes until golden and crisp. Sprinkle a little coconut over the top and serve. Serves 4.

AUGUST

8

Go Where Your Heart Draws You

August 1 - Planting the Seeds of Our Dreams

This day is celebrated by pagans as one of four fire festivals. Called Lammas, meaning "loaf mass," it marks the first harvest of the year, especially the first fields of ripened barley, wheat, or rye. The Celts believed that sacred energy transferred its power into the grain that remained in the field, so the last sheath cut was tied with a ribbon and ceremoniously kept until the following spring when the sacred grain was planted to ensure the success of the new year's crop.

Throughout life we continue to plant our own seeds—the seeds of our dreams and our ambitions. Our vision is the rich soil the seeds are planted in, and we watch over our crops with determination and patience. What do you hope to harvest this year? Will it be passionate and remarkable?

August 2 - Beaches

Our beaches are extremely tough environments and endure the battering, year after year, of winter storms. The strong ocean tides can shift tons of shingle and sand, even boulders, along the coastline. Yet here is another place we humans are damaging with our thoughtless ways.

We disturb the tiny creatures who live in the rock pools with our nets and fingers. Study them in situ, not in a bucket, and never remove living plants and creatures from their homes, not even temporarily. Replace any rocks you overturn on your search to watch crabs and urchins as these are hiding places and shelter. The spiral seashells you may want for your collection make ideal homes for hermit crabs so leave the larger ones behind.

Take your trash home with you since cans overflow and, along with a strong wind and scavenging seagulls and squirrels, cause trash and pollution to be strewn along the beach; unsightly yes, but also a serious danger to birds, fish, and other creatures who get trapped in netting, plastic, potato chip packets, and fishing lines.

When out walking along cliffs and sand dunes, don't tramp about like an elephant but stick to the paths to avoid adding to more erosion. At the seashore be mindful of being eco-friendly: leave it as you find it, if not nicer. Do your best not to disturb wildlife, including natural plants and, as it's said, "Leave behind only footprints."

August 3 - Summer Feasts

When you have a birthday or anniversary to celebrate this month, make it a beach, river, or lake picnic; summer's end comes all too quickly and, with a little effort, it'll be so special that your guests will talk about how clever and imaginative you are right through to fall.

You'll need shade, so choose a huge tree, or erect a tent or awning if on the beach, or put up lots of large umbrellas. Spread out colorful blankets or old comforters and add pillows and folding chairs.

Pack scrumptious but easy food in a large cooler on wheels, remember the cold drinks and a melon and peaches, oh, and slices of carrot cake! Take towels, sunhats, bathing suits, flip-flops, and lots of smiles. Leave behind your watch, your lipstick, responsibility, and stress.

Don't just sit there, play with the children . . . make sand castles, and paddle in the water. Have fun and be silly.

August 4 - Summertime and the Cooking is Easy

Who likes to cook in summer? Not me! But we do have to eat and raw foods only take us so far. Here are some ideas that pretty much cook themselves with little preparation:

Sour and Sweet Garbanzos

1	tablespoon olive oil
1	small onion, chopped
1	yellow bell pepper, seeded and diced
1	can (8 ounces) crushed pineapple in natural juice
¼	cup cider vinegar
1	tablespoon brown sugar
2	tablespoons ketchup
1	tablespoon ginger root, grated
1	teaspoon hot sauce
1	cup water
1	tablespoon cornstarch
3	cups cooked or canned garbanzo beans
4	cups cooked brown rice, served hot

Heat the oil in a large saucepan over medium heat. Stir in the onion and pepper and sauté for 5 minutes. Add the pineapple in its juice, the vinegar, sugar, ketchup, ginger, and hot sauce. In a cup, stir the water into the cornstarch to form a paste and then stir it into the pan. Stir in the garbanzo beans. Bring to a boil and stir. Lower heat, cover, and simmer for about 10 minutes until thickened and heated through. Serve over precooked hot brown rice. Serves 4.

Be the Change You Want to See in the World

August 5 - Easy Peasy Vegan Dish

Spicy Black-Eyed Peas

1	tablespoon oil
2	green onions, chopped
1	green chili, seeded and chopped
1	green pepper, seeded and chopped
1	teaspoon ground cumin
1	teaspoon ground coriander
2	teaspoons curry paste or curry powder
5	cups black-eyed peas, cooked and drained, or use frozen
3	cups soymilk
½	cup chopped fresh cilantro
4	cups cooked basmati rice

Heat the oil in a medium pan and sauté the onions, chili, green pepper and spices until vegetables are softened, about 5 minutes. Stir in the peas and heat through. Gradually stir in the milk. Bring to a boil. Lower the heat to a simmer, cover the pan, and cook gently for about 20 minutes. Stir in the cilantro. Serve with hot basmati rice and Indian bread. Serves 4.

August 6 - Melt In Your Mouth Delicious

Limey Bananas

4	firm bananas
1	tablespoon vegan margarine
4	tablespoons dark brown sugar

2   limes, cut in half

      flaked almonds for garnish

Peel the bananas and cut in half across, then lengthways. Melt the margarine in a large skillet and fry the bananas over medium heat until golden, about 5 minutes. Sprinkle the sugar over the top and continue to cook for another 5 minutes. The sugar will become syrupy. Transfer the bananas onto 4 plates, squeeze half a lime over the top of each, sprinkle with almonds, and serve immediately with soy ice cream or yogurt. Serves 4.

August 7 - Not-So-Happy Meals

If you are not vegan, or even vegetarian, or if you have children, I urge you to read Eric Schlosser's book, *Fast Food Nation*. The statistics in the book amazed me, especially the amounts of fast food sold, "In 1970, Americans spent about $6 billion on fast food; in 2000, they spent more than $110 billion."

Eric also goes into the fact that fast food is marketed profoundly towards children, and more and more children are becoming obese. There is also a section on the horrors of factory farming and more horrors at the slaughterhouses. He visits one of the largest in the country where he says, "About five thousand head of cattle enter it every day, single file, and leave in a different form." Not only a horror for the cattle but also for the mainly young, Latino women who barely earn a living processing these carcasses, standing shoulder to shoulder in this grisly hellhole.

Next time you feel like a happy meal, consider the unhappiness that went into making it. Read it and weep. I always wonder how otherwise normal, sensible people can continue to eat meat. I decided to ask some meat eaters I knew and here were their replies: "I always have." "I like the way it tastes and why would I want to just eat rabbit food every day?" "Meat's easy to cook after a long day at work; I haven't got time to soak beans. I slap a steak on the grill and it's done."

I've heard it all, yet more and more people are becoming vegan. And if you absolutely have to have the flavors of meat, there are a gazillion faux-meat products out there to fool any carnivore—pretend ribs and chickless breasts in barbeque sauce, Soysages, soy burgers you would never know weren't meat, even Tofurky for Thanksgiving. You can also buy soy ground for chili, tacos, and spaghetti sauce . . . the list is endless. And they are naturally low in fat and cholesterol and high in "meat-appeal."

Eating out is easy—Mexican, Italian, salad bars, or simply ask the chef to prepare something veggie for you . . . I've never had a problem. Just be polite and nice about it. And why should I care if anyone thinks I'm weird? Go ahead, Bite me! Dare to be different; dare to save an animal's life, and your own.

August 8 - Mega-Healthy Meals

On a hot summer's day there's nothing nicer than a lunch or dinner that uses the freshest of just-picked vegetables in a soup with fresh bread, crunchy croutons or breadsticks, and a colorful salad. Light, flavorful, and mega-healthy.

August 9 - The Taste of Spain

Here is my favorite soup from a vacation in Spain.

Andalusian Gazpacho

1	cup day-old bread (Italian or French), toasted and cubed
4	large organic vine-ripened tomatoes, roughly chopped
1	English cucumber, peeled and seeded
½	red pepper, seeded and chopped
½	small onion, chopped
1	clove garlic, chopped
1	tablespoon balsamic vinegar
2	tablespoons virgin olive oil
½	teaspoon salt (optional)
	freshly ground black pepper, to taste

Reserve ¼ cup each of the bread, tomatoes, cucumber, red pepper, and onion. Place the remaining ingredients in a blender or food processor. Blend until it's how you like it, from chunky to smooth. Chill for one hour. Check the seasoning, sprinkle with each of the reserved ingredients, and serve in chilled bowls. Serves 2-4.

August 10 - A Bean Medley for a Summer Day

Toucan Bean Salad with Lime Vinaigrette

1	can (15 ounces) garbanzo beans, rinsed and drained
1	can (15 ounces) red beans, rinsed and drained
1	small yellow summer squash, thinly sliced
2	small carrots, cut into fine julienne strips

1	small red onion, thinly sliced and rings separated
4	ounces Monterey Jack with jalapeno pepper cheese , or your favorite cheese, cubed
¼	cup virgin olive oil
1	teaspoon grated lime rind
¼	cup lime juice
2	tablespoons fresh cilantro, chopped
1	tablespoon water

In a pretty serving bowl, combine the beans, squash, carrots, onion, and cheese. In a screw-topped jar, shake together the rest of the ingredients to make the vinaigrette, then pour this over the bean mixture and toss well. Cover and chill for about an hour. Toss again before serving. Serves 4.

August 11 - Break the Chains

Tammy Sneath Grimes made a commitment to better things when she formed "Dogs Deserve Better," a voice for chained and penned dogs. See her Web site for more information at *www.dogsdeservebetter.com.*

August 12 - Be True to Yourself

Everyone should carefully observe which way his heart draws him, and then choose that way with all his strength.

—Hasidic saying

August 13 - Your Farm is Your Sanctuary

In 1986, Gene and Lorri Bauston found a living sheep abandoned on a stockyard "deadpile." They rescued the sheep, named her Hilda, and created Farm Sanctuary. Within ten years, Farm Sanctuary became the nation's largest farm animal rescue and protection organization. They now have a New York shelter and one in California, 100 miles north of Sacramento. Their Web site is www.farmsanctuary.org. Not only does their organization rescue thousands of farm animals each year, but they are also involved in groundbreaking campaigns and investigations into farm animal cruelty.

Rather than take the kids to a zoo this summer, why not visit a sanctuary if you have one near you, and support the people who care about rescuing animals and educating people on trying to put right the wrongs that humans impart on our furry and feathered sharers of this earth?

August 14 - The Whole Tomato

Have you been trying all the wonderful fresh tomatoes you can buy at the market now, instead of just the ubiquitous, evenly shaped, mass-produced ones? Or maybe you are growing your own, organic, heirloom varieties, are given some by friends, or buy them at a local Farmers Market?

Tomatoes are a subtropical fruit and do not like the cold; keeping them in the refrigerator impairs their flavor; warmth and sunlight are what they like best, since like other fruits they continue to ripen after they have been picked. Tomatoes contain no fat, are high in vitamins C and A, with traces of calcium and iron. They are so versatile; use them to excess.

August 15 - Antioxidants and Flavor Too!

Jamaican Farm Fresh Tomato Glut Salad

5	pounds of fresh ripe tomatoes, chopped
1	large red onion, diced or thinly sliced
2-3	cloves of crushed garlic
8	colorful sweet peppers—green, red, yellow, orange—seeded and sliced
3	jalapeno peppers, seeded and chopped (do not touch your eyes after or they will burn horribly)
	Jamaican hot sauce to taste
1	can (15 ounces) of black beans, drained and rinsed
2	cups or more of yellow corn, cooked

Toss all the ingredients in a large bowl. Serve with pasta and rice salad to the whole neighborhood . . . have a tomato party! Serves 12.

August 16 - Customized Aromatherapy

Most of the women I know like to smell lovely and fresh. In fact, scented body products date back to the Ancient Greeks, Indians, Egyptians, and Babylonians. But instead of buying a store-bought perfume, which may well have been tested on animals (and which goes on to smell nothing like it did at the mall and now you don't even like it but it cost $50), why not make up your own using essential oils diluted in carrier oil? (If you have a daughter, do this together for some great quality time.)

First decide on the kind of scent that appeals to you—herbal, flowery, woody, citrus, or exotic. Do you want just one scent, such as jasmine, ylang ylang, or rose, or do you want to be more creative and blend two or three to harmonize and add a different depth?

Other than lavender oil, always dilute the essential oil in a carrier oil such as sweet almond, jojoba, or my favorite, apricot kernel oil. Start with about 60 drops of carrier oil to 30 drops of essential oil or oils. Shake well and allow it to blend in a cool, dark place for about a week for the various scents to harmonize.

Your local health store should carry all the oils you need and may sell small bottles for blending and storage, or check the Internet for mail order and more information about the various oils.

August 17 - Scents for All Occasions

A few more uses for essential oils:

- To deter bugs from landing on your skin mix 30 drops of lavender to two tablespoons of carrier oil and dab on your pulse spots.

- For a room spray, add 4-8 drops of lavender, lemongrass, and peppermint oil to a small plant sprayer and top up with water.

- If you are bitten or stung by an insect use a drop or two of straight lavender oil. Lavender can also be used on your dog.

- For poison ivy, poison oak, and stinging nettles which cause raised welts and really nasty itching, wash the area with soap and water ASAP, then apply neat lavender oil, or a cold compress with two drops of lavender or chamomile oil to stop the irritation and help the swelling go down.

• Use a cold compress of lavender, peppermint, eucalyptus, and/or chamomile to cool and soothe heat exhaustion, sunstroke, sunburn, itching, prickly heat, and headaches. Keep in a baggie in the refrigerator for emergencies and fast relief, and to take on day trips, car journeys, or plane rides.

August 18 - Gentle Education

Whenever you serve a meal to family or friends who are not vegan or even vegetarian, explain without preaching or appearing superior why you choose not to eat meat, fish, or dairy . . . food without a face. Be prepared to answer any questions and have some leaflets at hand if they seem interested. Never pass up any opportunity to educate, especially when they are about to sample your food. Remember, peace begins in the kitchen.

AUG

August 19 - Thoreau Knows

I have no doubt that it is a part of the destiny of the human race, in its gradual improvement, to leave off eating animals, as surely as the savage tribes have left off eating each other when they came in contact with the more civilized.

—Henry David Thoreau

August 20 - Austrians for Animal Peace

Hoorah! I read in the newspaper that Austria passed an animal anticruelty law. Tigers, lions, and other wild animals are now banned from circus acts, dogs will no longer have their ears and tails cropped, puppies and kittens can no longer be kept in pet store windows, dogs are not to be chained up, wear choke collars, or be restrained by invisible electric fences that shock them if they try to cross the line. In addition, by the year 2020, chickens must be free of cages and allowed to scratch in farmyard dirt.

This is a great start to animal peace. Now would the rest of the world please take notice and follow suit?

August 21 - A Shared Universe

I went snorkeling and noticed how gently the fish welcomed us into their world—as compared to the violence with which we welcomed them into ours. I became vegetarian.

—**Syndee Brinkman**

August 22 - Wit and Wisdom

How are you at giving out advice and doling out the benefits of your wisdom and experience? Do you, or do you keep your thoughts to yourself?

When I lived in London I had a friend called Richard who was always throwing advice into his conversations; he did it in a witty, charming way, but you never knew if he was being serious or not so he never gave you the chance to be offended. He said it as if it were someone else's point of view and not his own!

Here are some of his greatest hits:

• Did you know the fire brigade is busier now that more people bathe by candlelight?

• Stop drinking alcohol the moment you can't say "succinct."

• Nose studs leave a hole.

• Never shave your eyebrows; they grow back in very laughable shapes.

• Do you eat raw garlic because you are tired of strange men chatting to you?

August 23 - Oxford Delight

Here is a soup recipe perfect for a summer evening. It was inspired by a bowl of soup I ordered while in the Cotswolds, near Oxford.

Cotswolds Chowder

1-2	tablespoons vegan margarine
2	small leeks, white parts only, thinly sliced
3	stalks of celery, sliced
1	medium potato, peeled and diced
¼	cup white basmati rice, rinsed
8	cups of plain soymilk, plus extra water if necessary
1	cup parsley, chopped
1	cup fresh dill, chopped
1	teaspoon white pepper
	instant mashed potato granules or powder, to thicken
	chopped chives and dill to garnish

Melt the margarine in a large pot. Stir in the leeks, celery, potato, and sauté them for a few minutes, stirring. Do not brown. Stir in the rice and soymilk. Bring to a boil then reduce the heat to a gentle simmer for about 20 minutes until the potatoes are softened, adding some water if it needs it. Stir in the parsley, dill, and white pepper. Stir well, bring back to a gentle boil and cook for about 5 minutes. Salt will spoil the flavor so do not add! If you like thicker chowder, stir in a tablespoon or two of instant potato granules. Do not overcook. Garnish each serving with a sprinkle of chopped chives and dill. Serve with crusty bread and a sprinkle of sunflower seeds. Serves 6.

August 24 - Wildlife in Danger

When I was a child back in the 60s, I collected tea cards. Tea didn't come in little square bags; it was loose leaf and came in oblong packets. Inside was a little picture card for children to collect. My most treasured set was a series of 49 cards entitled "Wildlife in Danger." There was a box in art class where you could swap your doubles; it was my first insight into wildlife. I would love to have those cards now to see which of the animals did not become extinct and made it through.

Habitat destruction—too many humans and too much greed—has doomed millions of creatures and plant species to extinction worldwide, leaving us a bleaker world. Only when we humans realize we are all in this world together, and this is not just our world, can all live in harmony.

Long before my tea cards were printed, indigenous people respected nature and were a part of it. Every major decision that Native Americans made, nature was considered: they did

Be the Change You Want to See in the World

not over hunt, destroy unnecessarily, or pollute. We must follow their example to transform our decaying world by changing our attitude about the wild.

August 25 - A Fountain of Transformation

The sound of running water can help transform your garden into a place of peace and tranquility. Even the smallest of areas can be turned over to water.

As long as it's watertight, almost any container can be transformed into a water feature. Choose a colorful shallow bowl or create a bubbling urn, or a pebble-based fountain that will look and sound delightful as the water trickles gently, yet is also safe for children. I recommend moving water so that mosquitoes are not attracted to the water and birds will love to take showers and flutter their feathers. My hummingbirds love the gentle spray from my fountain for their morning bath while still hovering, and they grab the tiny bugs that like it too.

Of course you'll want to be mindful of your water consumption, but most of these ponds and fountains simply recirculate the same water around and around. Check your local garden center or hardware store for ideas and add an area of serenity.

August 26 - Something to Whet the Appetite

By tradition, Greeks only drink alcohol if accompanied by mezedes—nibbles. The only rules for the food are that it is nutritious, appetizing, and unusual, and combines soft and hard, subtle, and bright, and fresh and mellow.

These same versatile, colorful, very flavorful dishes make a fantastic easy summer lunch or buffet to serve to your friends. Much of the preparation is done in advance and the food is served at room temperature. It's also a great way to eat communally as meze are not to be eaten with any fuss, just everyone dipping, scooping, and dunking raw vegetables such as cucumber, yellow, orange, and red peppers, celery, and the very necessary pita bread into a big bowl of lemony, garlicky hummous. (For vegetarians rather than vegans, you could also add Greek Feta cheese, rounds of fresh goats' cheese, and finish with fresh fruits drizzled with honey and served with dollops of thick Greek yogurt and some baklava.)

August 27 - Vegan Friendly

Here are some more vegan ideas. I ask you to serve the freshest of ingredients with the best olives and the most virgin Greek olive oil you can afford. Reproduce that wonderful Cretan feeling of friendship.

- Dishes of green and black olives, rinsed and marinated overnight in chopped fresh oregano, virgin olive oil and a squeeze of lemon juice.

- Bowls of green pistachio nuts.

- Trimmed green onions and radishes.

- Small red, orange, and yellow tomatoes.

Be the Change You Want to See in the World

August 28 - A Greek Recipe
Cretan Salad

½   pound orzo (pasta)

Dressing

3   tablespoons extra virgin olive oil
2   tablespoons red wine vinegar
¼   cup chopped fresh mint leaves
2   teaspoons fresh oregano, chopped
½   teaspoon freshly ground black pepper

Salad

2   tomatoes, diced
4   scallions, chopped
1   small cucumber, peeled, seeded, and diced
2   cloves garlic, crushed
1   cup pitted Greek kalamata black olives, quartered
¼   pound soy feta, crumbled
1   can (15 ounces) garbanzo beans, drained and rinsed

In a small saucepan, cook the pasta in boiling water for about 10 minutes, until al dente. Drain well. In a small bowl, whisk together the dressing ingredients to form a vinaigrette. In a large serving bowl, combine the orzo, tomatoes, scallions, cucumber, garlic, and black olives. Toss in the vinaigrette, feta, and garbanzos. Chill for 1 hour before serving. Serves 6.

Kali orexi—bon appetit!

August 29 - Hydration is the Key

Cystitis is a nightmare—drink water, drink water, drink water!

August 30 - Free and Easy

Let's all take the day off today, shall we?

August 31 - Be of Use

Think of where you would like to volunteer some time this Fall. Make some calls today and find ways to be of use.

SEPTEMBER

Honor and Respect the Creatures

September 1 - Sisters of the Earth

Nature has been for me, as long as I can remember, a source of solace, inspiration, adventure, and delight.

—**Lorraine Anderson**, author of *Sisters of the Earth*

September 2 - A Cup of Tisane?

Have you ever tried herbal tisanes? Not the tea bag variety, but fresh, organic herbs, spices, fruits, or flowers, infused in boiling water. They are called tisanes because they contain no tea, which comes from the leaves of *Camellia sinensis*, the tea plant.

Tisanes are old-fashioned medicinal remedies as well as refreshing drinks, being healthy to both body and spirit, and naturally contain no caffeine. Mint tisanes are the drink of choice in Middle Eastern and North African countries, served in glasses rather than cups. They help with the digestion, nervous headaches, colds, and sickness. Rosemary is used for headaches and insomnia, and chamomile is calming and aids sleep. You can buy dried organic blends such as blackberry leaf, rosehip, hibiscus, jasmine, and rose petal, or dry herbs yourself to store for later use—herbs such as lemon balm leaves, lavender blossom, or German chamomile flowers.

September 3 - Refresh Yourself

To make a refreshing pot of tisane bring cold, fresh water to a boil in a kettle. Rinse the pottery or china teapot with some of the boiling water to warm the pot. Follow the directions on the packet for how much of the dried leaves to use, usually two tablespoons, and let it steep in the boiled water. Due to the different aromatic oils in each herb, fruit, or flower, steeping time varies from 5-15 minutes so you'll have to experiment at first, but be careful it doesn't turn bitter tasting.

When serving tisanes to your guests use your prettiest cups and saucers and place sugar, honey, sweetener, fruit slices, or sprigs of mint, lavender, or lemon balm depending on the tisane flavor. Breathe in the deliciously fragrant vapor and enjoy.

September 4 - Fresh Inspiration

I am always inspired by fresh fruits and salads, and love to combine them for simple, healthy lunches. Choose from whatever is in season in your area:

• Peaches, raspberries, blueberries, sliced banana, and papaya, garnished with seedless grapes.

• Peaches and kiwi slices, mango, blueberries, and halved red grapes.

SEP

September 5 - Fruit with Every Meal

Include fruit in all your meals and be more imaginative at lunch than a PBJ!

September 6 - Not Just for Breakfast Anymore

Lunch Muffins

2 ½	cups wheat bran
1 ½	cups whole wheat flour
1	teaspoon cinnamon
½	teaspoon ground ginger
¼	teaspoon ground nutmeg
½	teaspoon salt
1 ½	cups apple juice
¼	cup molasses
2	tablespoons canola or sunflower oil
¾	cup carrots, shredded
½	cup sunflower seeds
½	cup raisins

Preheat the oven to 350°F. Lightly oil a muffin pan. In a large bowl, stir together the wheat bran, flour, spices, and salt. In a small bowl, whisk well the apple juice, molasses, and sunflower oil. Add the wet to the dry ingredients and stir well to combine. Add the carrots, seeds, and raisins and gently fold them into the muffin batter. Fill each of the prepared muffin cups ¾ full with the batter. Bake for 25-30 minutes or until an inserted toothpick comes out clean. Remove from the oven and allow to cool for several minutes before removing them from the pan. Serves 12.

September 7 - Yes, We Have Lots of Bananas

Bananas are the perfect healthy snacking food suitable for babies and small children, the elderly, people who are ill and, packaged in their own yellow skins for easy transportation, they are great power bars for athletes. Bananas are available all year and are full of potassium, so a banana a day may help to prevent high blood pressure and protect against atherosclerosis. Potassium may also counteract the increased urinary calcium loss caused by the high-salt diets typical of most Americans, thus helping to prevent bones from thinning out at a fast rate. They are also good sources of vitamins B2 and B6, vitamin C, magnesium, biotin, and fiber. Bananas have long been recognized for their antacid effects that protect against stomach ulcers and ulcer damage.

September 8 - Go Bananas!

Here's my favorite vegan banana cake. Go bananas!

Banana Walnut Cake

4	large ripe bananas
1	cup apple sauce or apple juice
1	tablespoon canola or sunflower oil
1	teaspoon vanilla extract
1 ¼	cups whole wheat flour
2	teaspoons baking powder
2	teaspoons baking soda
2	tablespoons cornstarch
1	tablespoon ground cinnamon
½	cup walnuts or pecans, chopped
2	tablespoons brown sugar

Preheat oven to 350°F. Lightly oil an 8x8 inch cake pan. In a large bowl, mash the bananas, and then, with a fork, beat in the apple sauce/juice, oil, and vanilla. In a smaller bowl, whisk the flour, baking powder, baking soda, cornstarch, and cinnamon. Tip these dry ingredients into the banana mix and fold together until just blended. Fold in the nuts. Dollop into the baking pan and spread evenly. Scatter the brown sugar over the cake batter. Bake for 40 minutes or until a toothpick inserted into the center comes out clean. Remove from the oven and leave to cool in the pan for about 10 minutes, then turn out onto a serving plate, sugar side up. Serves 9.

September 9 - Flour's Power

I am always very happy to reduce to a minimum the packaging on foods and prefer recyclable materials to be used. However, there is one item of food that leaves me fuming . . . flour! I buy whole wheat for cakes and pastry, self-rising for scones, special flour for my bread making but when I open the bag, flour goes everywhere and before I've finished the bag it's split. If I decant it to a storage container clouds of flour envelop me and makes the dog sneeze. Bags of flour have me floored.

September 10 - A Bird in the Hand

Outside my window, I have four bird feeders filled with mixed seed. Two hang from sturdy tree branches, which suits the finches, one is a platform feeder to entice a squirrel or two, the last is a pottery dish for the mourning doves. In the cooler months I also hang some of those prepacked seed bells for the myriad of finches. Vary the food you have available for your birds and you will attract different varieties depending on your area. Keep a bird book handy to identify your colorful visitors. Get your children interested too.

Just as important as food, I have four birdbaths in sight. I change the water regularly. These dishes vary in size and water depth for birds to drink and bathe in. I have found though that birds don't really like a slippery surface, so use unglazed dishes.

I also have hummingbird feeders where Anna's and Allen's hummers visit. I use one part white sugar to four parts boiled water, allow it to cool, and change it every day or so (so that the liquid doesn't ferment). You can also buy hummingbird food that is bright red to attract these tiny birds. Again, change regularly.

September 11 - Get Active, not Re-Active

If, after 9/11, you feel you would like to help people more and make a difference in a community then *www.globalvolunteers.org* may be able to help you.

If you are planning a trip abroad, the Web site will help you volunteer to improve the lives of orphaned and vulnerable children in Southern India, work on village development projects including teaching, building classrooms and dormitories in Tanzania, or work in Spain teaching English, or other projects in Greece, Jamaica, Mexico, Costa Rica, or Ecuador. All fees you pay are tax-deductible.

September 12 - For the Good of Us All

There is nothing like looking into the eyes of someone you are helping And knowing you are making their lives better.

—**Alexander Barton**, American volunteer

September 13 - One Woman Changes the World

Dr. Goodall was at the top of my list of people I wanted to meet. On September 13, 2002 I did just that at "An Evening With Dr. Jane Goodall" in Laguna Beach, California. Dr. Goodall is an incredible speaker and, despite all the horrible things she has both seen and experienced, she still maintains her *Reason for Hope*, as I have, since buying and reading her book of the same name.

She told us "We have to stop leaving all the decisions to the so-called decision makers, but take matters into our own hands, realize that each one of us makes a difference, and that if everyone who cares acts in a way that is ethical, thinks about ethical things, then the world would be changed very fast."

Dr. Goodall is a very positive lady. She is unrelenting on the theme of "hope." She is my inspiration in this world of greed and in my despair for the future of our planet. I read, "We have to overcome a feeling of helplessness. We have to hope." Let her be your inspiration too.

September 14 - Eight Glasses a Day

By now I've made my pitch for drinking lots of water. But perhaps like me you are torn about what water to drink. Half of me wants to be pampered by the ease and better taste of bottled water, the other half is bummed by the marketing gimmicks and trend of it all. Is fancy water worth the expense? It is if it encourages children to drink it rather than sugary sodas, or to drink the eight glasses a day for good health, or if the water from your faucet tastes like it came from the radiator of a '68 Chevy.

And the fact is, there is nothing wrong with our tap water: America has one of the safest water supplies in the world, and a good way to go to further improve the safety, and the flavor, is by adding water filters in your home. These can be under-the-sink systems, faucet-mounted filters, or pour-through products such as jugs and carafes containing a carbon-based filter system.

September 15 - Bone Up

In America, osteoporosis (a progressive disease that deteriorates bone mass, weakening bones and causing fractures, especially the hips, wrists, spine, and ribs, without any pain until a fracture occurs) affects millions of people, 80 percent of whom are women, costing our society billions in medical bills each year.

Clinical studies prove that a high intake of protein, especially animal protein, increases calcium excretion in the urine, calcium that is leached from the body. Dairy may be high in calcium, but it is also high in phosphorous, which binds to calcium, making it less absorbable. The kidneys have to work overtime to rid the body of this excess urea (protein) which requires minerals; these minerals are excreted along with the urea, and one of them is calcium: it is drawn out of the bones, leaving them porous and weak. When too much of this happens, you get osteoporosis, and it can develop at any age.

Yes, we all need protein, especially older people who tend to eat less food, but too much is also bad for us. A vegan diet is naturally low in protein, but if you consume enough calories from a varied diet, you will consume more than enough protein for health. A diverse, healthy diet and knowledge of calcium-rich foods will keep those vegan bones strong enough to take them to a rich old age.

September 16 - Vegans Eat Calcium Too

Calcium-rich vegan foods include: cooked kale and collard greens, baked beans, black-eyed peas, garbanzos, and other white beans, steamed broccoli, cauliflower, sesame seeds, and tahini (which is made from sesame seeds and is great blended into hummus). Choose calcium-enriched orange juice, cereals, and soy products such as soymilk and cheese, energy bars, and a big spoonful of blackstrap molasses. Vary these foods—don't think you can eat just one or two of your favorites all the time.

September 17 - Strong Mind, Strong Bones

Being a healthy weight, not drinking too much soda, coffee, or alcohol, not smoking cigarettes, staying active by walking and weight training, and keeping stress levels down will help maintain your bone strength.

Be the Change You Want to See in the World

September 18 - The Cat's Cradle

California is known, amongst other things, as the land of fruits and nuts, but in 1990 those "nuts" thankfully passed a law that designated the mountain lion a specially protected mammal and prohibited sport hunting of the cat, although they can still be shot when they pose a risk to humans, pets, livestock, and property.

This magnificent cat has other names: puma, cougar, panther, catamount, and painter with a Latin name of *Felis concolor*, literally, cat of one color. It used to be found all over the North American continent and throughout Central and South America, but sadly it is found no longer in most of Canada or the eastern U.S. due to habitat loss and hunting, the encroachment of humans building in its territory, and the disappearance of its favorite food, the white-tailed deer.

Also, from 1907 to 1963, California paid bounties on 12,452 cougars and in 1969 the cats were classed as a game animal and could be shot as a cruel trophy sport.

Educate yourself, and draw your own conclusions by checking out cougar statistics on *www.nps.gov* (National Parks), *www.cougarfund.org*, and for some incredible photographs of this majestic cat, go to *www.mountain-lions.org*.

September 19 - What's So Sporting About Murder?

We huff and puff when tigers and lions in India and Africa are killed, yet here we allow hunters to shoot our precious cats in the name of sport.

September 20 - A Dangerous Game

Do you know what Genetically Modified (GM) crops are all about? Right now, it's the hugest food experiment ever, and guess what? It's too much, too fast, too soon.

Defenders of this industry claim it holds the key to improving the nutritional quality of almost any food: I read of an experiment that would produce a GM potato that absorbs less fat when fried. Oh, puh-leese, millions are spent on an experiment to do that? Another claim is that GM foods will increase yields to feed a world where millions go hungry. According to ActionAid, "There is already one and a half times enough food to feed the world. The root causes of hunger are poverty and inequality. GM foods do little to tackle these issues and may exacerbate them."

In real life, and nothing to do with any study, Argentinian farmers have found that they have doubled the use of herbicides on their GM crops compared to conventional varieties. Another thing ecologists don't want to see happen.

I think it's pretty clear that the driving force behind GM agriculture is profit for the companies and countries that are developing GM crops, and has nothing to do with the interests of the single farmer, or world hunger. (1 percent of all GM research is aimed at the crops used by farmers in poorer countries.)

I urge you to educate yourself on all aspects of genetically modified foods. Don't be bamboozled by claims of being "vital to world hunger." It's much more complex than that.

Be the Change You Want to See in the World

Don't allow Dolly the genetically engineered sheep's wool to be pulled over your eyes.

September 21 - Research and Realize

Check your computer's search engine for differing opinions and facts, and then decide for yourself whether GM foods are safe or dangerous.

September 22 - Corporate Greed

There is an ongoing fight in Southeast India, in Kerala, where since 2002, the local people have been protesting against a Coca Cola plant that is contaminating their wells and causing a severe water shortage to the 1,000 people whose lives depend on clean, readily available water, and not on Coke.

One Indian claimed his well was full before the plant, and now he can barely pump enough to water his palms—his livelihood—where his harvest of coconuts has dwindled from 1,000 a month to 250.

Maybe large corporations believe they are helping the local people by providing work and industry; these people may be poor, but they have rights and must not be disregarded and demeaned.

This is happening in countries all over the world and in every industry, yet when these people desperately need our help in times of national disasters, or low-cost drugs to help epidemics like AIDS or typhoid, where are we? Probably sitting on our hands and our fat butts. This world is so full of injustice, and it needs to stop. Do your part.

September 23 - The Life Abundant

Fall is a peaceful time. The sun hangs lower in the sky, casting cooler shadows that melt into the warmer, brighter colors of the autumn. Virgo has changed to Libra to bring a harmonious balance and equilibrium to our Northern Hemisphere.

Autumn celebrates a maturity of the seasons and of the harvest; not only in terms of fruits and vegetables, but also a mellow time to look at our own lives, our desires, and our expectations to see if they have borne fruit. Is your life fruitful? Did your life grow and blossom this year, and your hopeful ideas evolve and mature?

September 24 - Embrace the Known and the Unknown

Appreciate your energy, and value and nurture the life around you: of plants, of animals, of the people you love, and of those you have never met.

September 25 - Be an Eco-Warrior

Feel free to explain to everyone that your choice of a money-saving, fuel efficient little car is due to an unselfish concern for the environment. Be an eco-warrior.

September 26 - Horse Sense

The time is always right to do what is right.

—**Martin Luther King, Jr.**

September 27 - Early Harvest Treats

Make the most of the glut of early fall vegetables in local markets now. Here is one of my favorite recipes:

Stuffed Peppers

6	whole bell peppers (red, yellow, orange, or mixed)
1/2	red bell pepper, seeded and chopped
1	small onion, chopped finely
1	zucchini, chopped
1	yellow squash, chopped
1	tablespoon olive oil
1 1/2	cups cooked couscous or brown rice
1/3	cup fresh Italian parsley, chopped
1	tablespoon thyme leaves
1	tablespoon lime juice
1	tablespoon balsamic vinegar
	freshly ground black pepper, to taste

Slice the tops off the 6 peppers, keeping them whole. Core an seed them and cut a little off their bottoms so they can stand up, but don't slice too deep or the filling will fall out! In a small pan, sauté the red pepper, onion, zucchini, and squash in the oil until softened—about 10 minutes. Remove from the heat. Stir in the cooked couscous or rice and the parsley, thyme, lime juice, balsamic vinegar, and black pepper. Taste to check seasoning. Spoon into the pepper shells and serve straight away, or cool and serve at room temperature. Serves 6.

September 28 - Nutritious and Delicious

Ratatouille Provencal

1	large onion, chopped
2-3	cloves garlic, crushed
2	bell peppers (red and green), seeded and sliced
2-3	zucchini, sliced thickly
1	small eggplant, cut into chunks
1	tablespoon olive oil
1	tablespoon dried oregano
2-3	vine-ripened tomatoes, diced
½	small can (4 ounces) tomato paste
1	cup vegetable stock
½	cup fresh parsley, chopped
½	cup fresh oregano, chopped
	salt and freshly ground black pepper, to taste
	French bread

In a large pan, sauté the onion, garlic, peppers, zucchini, and eggplant in the olive oil for about 10 minutes until it smells wonderful! Stir in the dried oregano and tomatoes. Combine the tomato paste and stock, and then stir this gently into the stew. Simmer until cooked to your liking, about 30 minutes or so. Gently stir in the parsley and oregano, season, and serve with crusty French bread. Some people also like it cold the next day. Serves 6.

Be the Change You Want to See in the World

September 29 - Everything Nice

Autumn Fruit Pudding

4	cups mixed blackberries, chopped apples, and chopped pears
¾	cup light brown sugar
1	teaspoon ground cinnamon
7	tablespoons springwater
	about 12 slices thinly sliced white bread, crusts removed

Place the fruit in a pan with the sugar, cinnamon, and 7 tablespoons of springwater, stir, and bring to a boil. Reduce the heat and simmer gently for 5-10 minutes so that the fruit softens yet holds its shape. Line the base and sides of a 4 cup pudding bowl with the bread slices—be sure there are no gaps between the slices anywhere (and save enough bread to cover the top). Spoon the fruit into the bread-lined bowl and top with the remaining bread. Put a saucer down on top of the bread and weigh it down with some cans or something. Place on a plate to catch any drips and chill in the fridge for at least 8 hours. Turn the pudding out onto a serving platter and serve immediately with soy ice cream. Serves 4-6.

September 30 - Brother Chimp, Sister Chimp

Did you know it is legal in the U.S. to keep an adult chimp alone in a cage that is 5 feet, by 5 feet, by 7 feet for its entire life?

Many chimpanzees in the United States devote their formative years to biomedical research, the entertainment field, or parks/zoos, and then when they are no longer wanted, get sick, or too old, euthanasia is the next step. No chimpanzee ever volunteered to be confined by humans for selfish purposes so we must ensure that their contribution to humans is rewarded by providing them with the best possible care for the remainder of their years (which can reach beyond 50 years).

In San Antonio, Texas, Primarily Primates, Inc., founded in 1978, approved of by the ASPCA and Dr. Jane Goodall, now cares for a community of more than 600 primates, 150 birds, and various other mammals. It is a unique nonprofit animal protection organization that provides sanctuary, rehabilitation, lifetime care, and shelter to animals that otherwise would continue to suffer from abuse, neglect, or be killed as unwanted surplus.

It really is morally unacceptable to use, abuse, and then murder any animal but especially our sibling species. Educate yourself further on primates at *www.primarilyprimates.org* and do what you can to help these precious creatures.

OCTOBER

Harmonious Balance

October 1 - An Ounce of Prevention

October is National Breast Cancer Awareness month and every human owes it to himself or herself to be educated about this form of cancer, indeed all forms of cancer.

Breastcancer.org is a nonprofit organization for breast cancer education and a good place to start with any questions you may have. Before you make your next appointment for a mammogram, educate yourself on breast cancer and cancer cells.

October 2 - Vegan Breast Cancer Prevention

Prevention is helped greatly by a vegan diet which is rich in soy foods, legumes, green leafy vegetables, citrus fruits, and is low fat. Foods such as soymilk, tofu, and tempeh contain hormonelike substances called phytoestrogens and research suggests these help block the breast cancer-inducing effects of the estrogen hormone. Then add to this an active lifestyle in the fresh air, and easing off the alcohol and excessive stress.

October 3 - Dish it Up

Smothered and Squashed Red Beans

1	tablespoon olive oil
½	cup chopped onion
1	jalapeno pepper, seeded and chopped
2	garlic cloves, crushed
1	cup yellow squash, sliced ½ inch thick
1	cup zucchini, sliced ½ inch thick
1	cup frozen or freshly cooked corn kernels

1   can (15 ounces) red kidney beans, drained and rinsed

1   can (15 ounces) diced tomatoes, undrained

3   thyme sprigs

     salt and freshly ground black pepper, to taste

2   cups hot, cooked long-grain rice

½   cup (2 ounces) shredded vegan cheese

Heat the oil in a large skillet over a medium-high heat. Stir in the onion, jalapeno, and garlic, and sauté for 2 minutes. Stir in the squash and zucchini, and sauté for a couple of minutes. Stir in the corn, beans, tomatoes, and thyme; cover, reduce heat, and simmer for 10 minutes or until everything is cooked nicely. Season to taste. Discard the thyme sprigs, serve over rice, and sprinkle with cheese. Serves 4.

October 4 - Spicy, Savory, and Meat Free!

Spicy Senegal Stew

1   tablespoon peanut or canola oil

1   large yellow onion, chopped

¼   white cabbage, chopped

2   cloves garlic, crushed

1   can (18 ounces) sweet potatoes, drained and chopped
     (or use freshly cooked and peeled)

1   can (15 ounces) diced tomatoes

2   cups tomato juice

1   cup apple juice

1   tablespoon ginger root, grated

1   tablespoon ground cumin

OCT

1   teaspoon red pepper flakes
2   cups frozen edamame (green soy) beans, defrosted
1/3 cup peanut butter

In a large pan over a medium-high heat, sauté the onion in oil for 5 minutes. Stir in the cabbage and garlic and sauté for a further 5 minutes. Stir in the sweet potatoes, tomatoes, tomato juice, apple juice, ginger, cumin, and red pepper flakes. Reduce the heat to low, cover, and simmer for 5 minutes. Stir in the beans, leave uncovered, and simmer for another 5 minutes. Stir in the peanut butter and heat through for about a minute until the stew is hot and bubbling. Serves 4-6.

October 5 - Autumn Baking

Baked Apples

4   large Granny Smith apples, cored
1   cup good quality maple syrup
1   cup pecans, chopped

Preheat the oven to 350°F. Place the cored apples into a lightly margarined baking dish. Pour the syrup over the apples, sprinkle with nuts, and bake for 35 minutes. Great with soy ice cream! Serves 4.

October 6 - A Friend for Life

Why not make fall the time you schedule your breast examination check-ups or mammograms? Ask a close friend or female relative to have hers at the same time for emotional support. Remember, the two greatest risk factors of breast cancer are being a woman and growing older. Survival rates are high once breast cancer has been diagnosed.

If someone you know is diagnosed with any cancer, intense emotions will surface: emotions that range from anger, to fear, to sadness, and back again. There is no correct response to these emotions, but you should allow your friend to honestly express how she is feeling. Talk to her, ask her how she is feeling, what she would like from you, and what she needs; also ask her what she doesn't want. Please listen, really listen.

October 7 - In Sickness and in Health

Each person copes in her own way with the stress and fear. Your love and support will provide the emotional space a friend needs as she works through this thicket of feelings, allowing her to deal with her grief in her own way.

October 8 - Nurture Creativity

Children must use all their senses for full cognitive and academic development. If you have children, grandchildren, or a niece or nephew, encourage them to act out their fantasies, play at dress-up, painting, reading, and being creative. Sit down with them, join in too, talk to them, and be sure you listen to their replies, encourage their language skills which will develop into expressive social skills. Childhood years are too short; let children enjoy them.

OCT

October 9 - An Eye for an Eye Makes the World Go Blind

There are so many simple but meaningful ways to help out in this world.

For instance, I started bugging my friends for their eyeglasses. No, it's okay, I have some of my own, but I want their old ones. I want them to sort through their drawers for glasses they no longer need. Why? Lions Clubs and other organizations all over the U.S. collect these no longer wanted spectacles, clean them, repair them, and classify them by prescription, then deliver them to people in developing countries who so desperately need glasses but can't afford them.

Go online to find details of a group you could support. I contacted United For Sight, who send donated glasses to Somalia, Uganda, Congo, Cameroon, Tanzania, Kenya, Guinea, Ghana, Benin, India, and Thailand. They suggest you can begin the eyeglass drive by obtaining plastic bins and labeling them with posters. Place the bins in several prime locations, including offices, libraries, eye doctor offices, and hospitals. It's also a great school or scout group project.

If you need to wear glasses for reading, driving, or even so that your loved ones' faces are no longer a blur, you'll know how miserable life would be without them. Recycling eyeglasses you no longer need is a great—and absolutely painlessly simple—thing to do.

October 10 - Fur is Murder

Millions of animals suffer and die every year to provide humans with fur coats, hats, and other accessories. It doesn't matter whether these animals were farmed or lived in the wild—it makes no sense to clothe yourself in a pelt that looked so much better on its original owner. Why would you want to drape a skinned fox around your shoulders, head and tail still attached?

Visit www.banfur.com, the coalition to abolish the fur trade, where you can read the horrors that lynx, beaver, otters, foxes, raccoons, mink, coyote, and bobcats suffer in the name of fashion and one-upmanship—and they will give you info on what you can do.

October 11 - Walk Your Way to Health

Walking is one of the simplest and safest forms of aerobic exercise you can do. It will help strengthen your bones, your heart, and your lungs, as well as controlling your weight if you do it regularly and briskly. Walking is also easier on the knees, hips, and ankles than running or jogging. People who walk 20-25 miles a week tend to be healthier and outlive, by several years, those who don't exercise. The elderly too benefit greatly by walking, by keeping them mobile, independent, and happier.

Treat yourself to a new pair of ladies' walking shoes, wear warm, cotton clothing in layers so that you can remove an upper layer as you get warmer to tie around your waist, warm up with a slower walk first, put a smile on your face,

OCT

stride out, enjoy your walk, and be sure to cool down and stretch at the end.

October 12 - Animal Rights Abroad

Here's another tragic fact about our friends from the wild. In Namibia, in southwest Africa, over 7,000 cheetahs were killed between 1980 and 1990 by farmers who saw them as a major threat to their livestock. Namibia had gone through a terrible drought in the 1980s, which killed off the cheetah's natural prey, so the cheetah had turned to the farmers' cattle as food.

There are only about 12,000 cheetahs left in the wild, in places such as Zimbabwe, Kenya, South Africa, and Iran . . . in Namibia only about 2,500 remain, where they are a protected species now, yet farmers are still allowed to "remove" a cheetah by trapping or shooting it if it poses a threat to their livestock.

I went online to *www.cheetah.org* to read more about the Cheetah Conservation Fund's work where they are researching and implementing strategies to save the cheetah in its natural habitat, and working with the people who also share the cheetahs' home. The part that fascinated me was the brilliant idea of using Anatolian Shepherd dogs to guard the livestock. These dogs are huge, around 155 pounds, and come from Turkey and the arid region of Asia Minor, a climate similar to cheetah country. They are introduced as puppies to the livestock herd and they bond with them, eat and sleep with them, and basically protect them, not only from cheetahs but also from baboons, jackals, leopards, and even humans—they bark up a storm and frighten away any animal that comes too close. Cheetahs are not aggressive naturally so they quickly find another food source. The dog saves the livestock . . . and saves the cheetah from a bullet.

If you love the world's fastest land animal and want to know more, go to this Web site. You can sponsor a cheetah cub, an adult cat, or help pay for an Anatolian guard dog—all save the cheetah. There are some great photographs of cheetahs and the Anatolian Shepherd dogs on the site, so check it out!

October 13 - A Loaf of Compassion

Bread plays a very important part in a vegan diet. A fresh loaf of whole wheat, or whole grain, bread makes an instant nutritious meal with a generous spread of peanut or almond or cashew nut butter and homemade jam. Check the bread label, though, for any milk products, or even honey, if you are a strict vegan. Better yet, bake your own.

Bread is also great with soups, with a baked bean topping, made into breadcrumbs, or used to mop up those last tasty morsels of a veggie stew.

October 14 - Fast Food Nation

There are estimated to be over 20 billion head of livestock on our planet, that's triple the number of people. U.S. pork and beef consumption has tripled since 1970 and since 1961 the number of fowl raised globally for consumption has quadrupled.

One reason for this huge increase is the rise of fast-food restaurants, especially here in the U.S. In his book, *Fast Food Nation*, Eric Schlosser writes, "Americans now spend more money on fast food—$110 billion a year—than they do on higher education. They spend more on fast food than on movies, books, magazines, newspapers, videos, and recorded music—combined."

Don't be brainwashed. We all know how unhealthy these meals are, unhealthy for our bodies, and unhealthy for our planet. Eat vegan, and don't be bullied by burger joints.

October 15 - The Little Things That Count

Back in the 80s my partner, Stephen, and I took a trip to northeastern Spain, just across the border with France. We stayed in a little fishing village called Rosas, which is at the northern part of the Costa Brava. We stayed in a simple room above a lively bar called Pete's Place, which the locals used on their way home from work. The TV in the back bar showed Spanish football and old Clint Eastwood films dubbed in Spanish!

Each evening, after a long day's sightseeing, we'd join the locals for a glass of red wine, or Sangria, and enjoy the rustic food called tapas, served at the bar on rustic plates and bowls.

Tapas originated in Andalusia, a region of southern Spain, where the people gather in the warm evenings to enjoy lively conversation, a glass of golden fino sherry, and tapas. Tapas are a great way to serve food to your friends . . . a range of mouthwatering, fun, happy hour finger food . . . simple to put together, yet with a Spanish theme. Many markets offer

Spanish wines and sherry; so take it from there. Here are some ideas to serve on your chunkiest plates and bowls and wooden platters: Roasted tiny red-skinned potatoes, marinated mushrooms and Spanish green olives spiked on toothpicks and sprinkled with olive oil and thyme leaves.

October 16 - Try These Tapas

Pine Nut-Stuffed Mushrooms

16	large button mushrooms
1	tablespoon virgin olive oil
1/2	small onion, finely chopped
1	celery stalk, finely chopped
1/2	cup pine nuts
1	clove garlic, crushed
1	tablespoon fresh thyme leaves
1/3	cup breadcrumbs
2	tablespoons freshly chopped parsley
	salt and freshly ground black pepper, to taste
4	slices of vegan cheese, cut into quarters

Lightly oil a large baking dish and set aside. Remove the stems from the mushrooms, roughly dice the stems, and set aside. In a nonstick skillet, sauté the mushroom caps in oil for 2 minutes per side. Remove the mushroom caps from the skillet, transfer to the baking dish, and set aside. In the same nonstick skillet, sauté the reserved mushroom stems, onion, and celery for 5 minutes to soften. Add the pine nuts, garlic, and thyme, and sauté an additional 2-3 minutes until the vegetables are tender. Remove the skillet from the heat, add the breadcrumbs and parsley, stirring well to combine, and season to taste. Fill the mushroom caps with the warm filling, slightly mounding to use it all. Top with a piece of sliced cheese. Bake at 350°F for 5-10 minutes or until the mushroom caps are heated through and slightly browned on top. Transfer to a decorative platter and serve immediately. Serves 6-8.

October 17 - A Spanish Delight

White Bean Paté

6	cups water or vegetable stock
1	pound dried lima beans, sorted and rinsed, and soaked for 6-8 hours
2	bay leaves
¼	cup freshly chopped parsley
1	tablespoon virgin olive oil
	salt to taste and lots of freshly ground black pepper

In a large pot, combine the water, beans, and bay leaves, and bring to a boil. Reduce the heat to low, cover, and simmer until tender, about 45 minutes. Check a few times to make sure the beans don't stick and burn. Remove the pot of beans from the heat and discard the bay leaves. Using an immersion blender or food processor, process the beans and their cooking liquid to form a smooth puree. Add the parsley, oil, salt, and pepper and process a few times to combine. Taste and adjust the seasonings, if needed. Serve with crusty bread. Serves 6-8.

October 18 - I'll Drink to That

Spanish Sangria

4	cups dry red wine
1	cup club soda
½	cup each orange and cranberry juice and orange liqueur
¼	cup superfine sugar
	ice cubes
2	oranges, sliced

Combine wine, club soda, juices, and orange liqueur in a punch bowl or large pitcher. Stir in sugar until dissolved. Add ice cubes and orange slices. Serves 8.

October 19 - Keep Your Germs to Yourself!

As we head into winter, the sniffles begin. In fact the most common illness in the world is the common cold. In the U.S., 90 percent of us will have at least one cold this year. Colds are caused by a number of different viruses; this is the reason there is no "cure." These germs attack the moist skin lining our noses, sinuses, throat, and upper respiratory tubes, causing sneezing, runny nose, coughs, sore throat, and that horrible sinus congestion.

The way you catch a cold is via other people, spread by tiny droplets in the air by a sneeze or a cough, and from direct contact: by using the phone, or handshakes, etc. Infected people are able to spread their germs a day or two even before their symptoms start to show, and then for three to four days

OCT

after. So, if you have a cold or the flu, stay at home if you can, rest, and keep those germs to yourself. If you really can't stay home, then wash your hands frequently, flush your used tissues in the toilet, and cover your mouth when you cough or sneeze. You can buy antiseptic wipes to clean your phone. If you smoke, give it up, and if you have a humidifier at home this will help keep air moist and help ease your symptoms. Use aroma oils such as eucalyptus and peppermint on a tissue to help you breathe, and lavender to soothe and help kill germs.

October 20 - Pamper Yourself

Pamper your body's immune system to keep it strong to fight infection before it can take a hold. Eat healthily—lots of raw vegetables and fruits. Keep a bag of cut vegetables ready in the fridge each morning to munch on, eat apples, berry fruits, drink fresh juices instead of cola or coffee, and have a big crunchy salad for lunch and dinner rather than a pizza or fast food.

Get plenty of sleep each night and plenty of fresh air throughout the day by taking a walk, maybe at lunchtime, to feel revitalized and positive. Associate more with upbeat people and laugh. If you start to feel negative, call someone who will cheer you up and make you feel good again.

Work at having a healthy mind and body, keep active, destress regularly, and when your body feels tired, listen to it and rest. Pamper yourself; your body will love you for it.

October 21 - The Magic of Pumpkins

Don't you love visiting the market in fall and seeing all those glorious orangey gold pumpkins? The pumpkin is one of many winter squashes available, full of vitamin A, B vitamins, fiber, iron, and potassium. For cooking, choose a sugar pumpkin, about 5-6 pounds in weight (smaller and sweeter than the Jack-o'-lantern variety), cut off the top, an inch or so below the stalk, and scrape out the seeds and stringy pulp with a spoon. Bake in a 350°F oven, the cut side down on a foil-lined baking pan, for an hour or so. The pumpkin should be tender yet not so soft that it collapses. Place in a serving bowl, carefully pour in your soup, and, as you serve into individual bowls, scoop some of the pumpkin from the shell, being careful that you don't break through the shell.

October 22 - Squash Gone Wild

Stuffed Squash Gone Wild

2	medium acorn squash
1	cup fresh cranberries
1/2	cup brown sugar
1/4	cup sliced almonds
1	cup cooked brown rice (¼ cup precooked)
2	cups cooked wild rice (½ cup precooked)
1	celery stalk, chopped
1/3	cup freshly squeezed orange juice
2	tablespoons apple cider vinegar

Preheat oven to 350°F. Halve the squash lengthways and remove the seeds and stringy bits. Place the squash face down in a baking dish and bake, uncovered, for 30 minutes or until tender. In a medium pan, cook the cranberries, sugar, and almonds for about 5 minutes over medium heat until the sugar melts—stir well to prevent burning. Stir in the cooked rice, celery, orange juice, and vinegar. Heat through for about 5 minutes. Turn the squash so the cut side faces up, and spoon the mixture into the baked squash. Serves 4.

October 23 - Make Room at the Inn

With the colder, wetter weather approaching please keep an eye out for local cats and dogs. If you see anything mean or any obvious neglect, contact the authorities—your local animal shelter, humane society, or sheriff's department.

October 24 - Don't Start That Car

Cats will often curl up under still-warm cars for warmth. Always take a quick check before your start your car's engine, especially if you know there are outdoor cats in your area. Be sure to use an antifreeze that is relatively non-toxic to animals, such as Sierra brand. Store it in a safe place away from animals, and if you spill any or it leaks, clean it up immediately. Antifreeze tastes sweet and is extremely poisonous if lapped by an animal. For more cold weather tips and details on antifreeze, go to *www.aspca.org*.

October 25 - Consider Child Safety

If you have children, grandchildren, nieces, nephews, are a teacher, or are in any other way connected with children, educate yourself, and those children, about the Amber Alert. The Amber Alert was created in 1996 to raise awareness for child safety as a legacy to little Amber Hagerman who, at age nine, was kidnapped and murdered in Arlington, Texas. Their Web site at *www.codeamber.org* has lots of information on how you can help, but the thing I liked best was their instructional video, called *Safety Net Kids*, recommended for ages 4-11 and designed to watch again and again to reinforce the lessons. You can buy it online.

According to the National Center for the Missing and Exploited Children there will be over 58,000 nonfamily abductions this year . . . educate yourself and your family so that your child isn't one of them.

October 26 - South of the Border, Veggie Style

Chili started south of the border, but it's become an American staple. Check out any veggie cookbook and you'll probably find several different recipes . . . red beans, pinto, black, or a mixture of all three, with tomatoes, without, fiery hot or mild, with fresh chilies or with cayenne. Then there's what to serve it with: rice, tortillas, baked jacket potato, quinoa, guacamole, sour cream. It's your choice, just make sure it's healthy, has lots of flavor, tingles your taste buds . . . and there's enough left for tomorrow! Here's one of my favorites:

OCT

Lonely Cowboy Chili

1	package (12 ounces) firm tofu, frozen, defrosted and drained
1	tablespoon olive oil
1	large onion, chopped
2	cloves garlic, crushed
2	large red and/or green peppers, seeded and sliced
2	jalapeno peppers, seeded and chopped
3	tablespoons chili powder
1	tablespoon dried oregano
1-2	tablespoons ground cumin
1	can (15 ounces) red kidney beans, drained and rinsed
1	can (36 ounces) crushed tomatoes
5	cups vegetable stock
½	small can (4 ounces) tomato paste
2	cups frozen corn
½	cup fresh cilantro, chopped
½	cup fresh parsley, chopped
	salt and freshly ground black pepper, to taste

Freezing and defrosting tofu gives it a firm, almost spongy texture–try it. Cube the tofu and, in a large pan, sauté it in oil with the onion, garlic, peppers, jalapeno, and seasonings for about 15 minutes, stirring occasionally so that nothing burns. Add the beans, tomatoes, and stock, bring to a boil, cover, and lower the heat to a simmer for about 45 minutes. Stir in the tomato paste, corn, and fresh herbs. Heat through for a further 15 minutes or so, taste, season, taste again–make it perfect! Serves 6.

October 27 - Are You Chilly? Eat Some Chili!

And here's another:

Not a Texas Chili!

4	veggie burgers
1	tablespoon olive oil
1	large yellow onion, chopped
1	cup water
1	can (28 ounces) diced tomatoes
1	can (15 ounces) red kidney beans, rinsed and drained
1	small can (4 ounces) tomato paste
1-2	jalapeno peppers, seeded and chopped
1	tablespoon chili powder
2	teaspoons brown sugar
2	teaspoons unsweetened cocoa powder
1	tablespoon dried oregano
1	tablespoon ground cumin

In a medium nonstick pan, cook the veggie burgers as directed on the package and break them up with a wooden spoon as they begin to brown. Stir in the oil and onion and sauté for about 5 minutes. Stir in a cup of water and the tomatoes, beans, tomato paste, jalapeno peppers, chili powder, brown sugar, cocoa, oregano, and cumin. Bring to a boil, cover, and simmer for 15-30 minutes. Check for seasoning. Serves 4-6.

October 28 - The Organic Apple of Your Eye

We're in the thick of apple season now. Wonderful varieties are grown from coast to coast. Organic apples are especially good for you and the environment as no chemicals or pesticides are sprayed on them. Apples contain vitamins A and C, potassium, iron, calcium, and its real benefit, fiber—the fiber in the skin plus pectin, a soluble fiber which binds to cholesterol and bile acids and helps secrete them from the body. Apples contain only 80 calories, and make a great instant snack, straight from the fridge or fruit bowl.

Apples also stimulate our digestive juices, they are detoxifying, and help protein digestion. They keep our digestive tract healthy and prevent constipation. For these reasons eat apples au naturel—unpeeled. This is also where the nutrients are.

Visit a farm stand, or pick your own apples this month, for yourself and to give away as fruit, or as a pie—do your bit to keep the doctor away!

October 29 - Radical Self-Care for Women

Women still don't make time for ourselves! We've gained an "extra" hour this week. Use that as a reminder to make time for yourself this autumn, to do things for yourself, and for your soul. And, just as importantly, never feel guilty about it.

October 30 - Making Halloween Mishchief

Got your candy all ready for tomorrow night? Pumpkins all carved and glowing with night-lights, fake witch splattered against the garage door, and are the children's costumes all ready for them to wear?

Such harmless fun and mischief, but how about if you have no kids? Throw a party of your own, and get your adult guests to dress up.

Vegan party food? Greet your guests with apple cider, spiked or not with brandy, kept warm in your slow cooker. Black olives stuffed with blanched almonds to resemble eyes, cocktail onions or tiny tomatoes and cubes of vegan cheese spiked onto cocktail sticks (for a burst of heat I like to use cheese with jalapeno peppers), herbed bean paté with sticks of raw peppers, cucumber, and baby carrots to dip. As a main dish, made earlier in the day and reheated for your guests, serve a warming Cottage Pie, full of seasonal vegetables, and for dessert your favorite home baked cake, maybe with ginger and spices.

As an old Victorian tradition wrap tiny charms in waxed paper and insert into the cake—a coin means wealth, a wishbone means wishes and dreams, a heart means love, a ring means marriage, a key means success, a button means bachelor, and a thimble means spinster.

Halloween isn't just fun for kids!

October 31 - All Hallow's Eve Party Pie

Cottage Pie

2	tablespoons olive oil
1	onion, chopped
8	ounces mushrooms, chopped

2	sticks celery, chopped
2	carrots, sliced
1	cup green lentils, rinsed
1	tablespoon tomato paste
1	cube vegetable bullion
	salt and freshly ground black pepper, to taste
1	tablespoon dried oregano
1	tablespoon fresh parsley
2	bay leaves
2	cups white beans, cooked
½	cup good red wine (optional)

Topping

3-4	large potatoes, cooked and mashed with soymilk
1-2	tomatoes, sliced
	vegan cheese, grated

Preheat oven to 375°F. In a large pan, sauté in oil the onion, mushrooms, celery, and carrots for about 10 minutes. Add the lentils, tomato paste, bullion, salt, pepper, oregano, parsley, bay leaves, beans, and wine. Bring to a boil, adding more water to cover if necessary. Lower the heat to a simmer and cook, covered, for about 30 minutes until the lentils are softened. Check the seasoning. Place the veggie mixture in a large ovenproof dish, top with the mashed potatoes, and form into peaks with a fork. Place the tomato slices decoratively on top, sprinkle with cheese, and bake for about 40-45 minutes until golden and bubbling. I place the dish on a pan lined with foil to catch any drips. Serves 6-8.

NOVEMBER

Appreciation

November 1 - The Month of Reflection

November

The mellow year is hastening to its close;

The little birds have almost sung their last,

Their small notes twitter in the dreary blast—

That shrill-piped harbinger of early snows;

The patient beauty of the scentless rose,

Oft with the morn's hoar crystal quaintly glassed;

Hangs, a pale mourner for the summer past,

And makes a little summer where it grows:

In the chill sunbeam of the faint brief day

The dusky waters shudder as they shine;

The russet leaves obstruct the straggling way

Of oozy brooks, which no deep banks define,

And the gaunt woods, in ragged scant array

Wrap their old limbs with somber ivy-twine.

—**Hartley Coleridge**

November 2 - Walking the Weight Off

Once I worked as a physical instructress to a Saudi Arabian princess, who had gained weight following an accident to her ankle. Her physician had told her she was becoming too heavy and should exercise, so I was hired. Unfortunately, when you are a very wealthy princess and merely have to clap your hands for everything you want, losing weight proves to be very difficult, but I tried to encourage her.

She wanted to lose inches from her stomach, but, being a princess, would only do sit-ups on the bed, not the floor; she didn't want to walk as it was too hot, she didn't like swimming, and she found it too difficult to cut down on calories.

From Riyadh we moved to one of the finest health spas in Europe, in Evian, France, just over the Swiss border. The princess felt massage would help—she had at least one a day, but she never walked the grounds, played tennis, or swam. She was advised to eat the special meals in the restaurant that were healthy and lower in fat, but she liked them so much, she ate twice as much!

After a month we all moved on to her summer home in Beverly Hills. I equipped her pool room with weights, but she trained only once a week. After a month or two, I coaxed her into the shallow end of the pool for aqua-aerobics. She did well and I thought we would now make progress . . . but I wasn't summoned again until a week later; her servants told me she had discovered cheesecake, and they had yet to try any leftovers! Several more times I met her at the guest pool, but it was me who ended up swimming laps while she watched from the lounge chair and burned calories talking on the phone to her friends in Riyadh.

Towards the end of the summer she joined me for aqua-aerobics again. She did some exercises using the weights and I was highly encouraged that finally I could do the job I was here for. Yet as I walked up the long driveway back to the main house, she was sending her bodyguard out for a double burger, fries . . . and a large diet Coke. I quit, totally discouraged, and took a flight back to England. I couldn't motivate her; I felt I had failed.

Looking back, I realize she didn't want to lose weight; it was her doctor who thought she should. She was rich enough to send someone out to buy a new closetful of clothes in the next size whenever she chose, and she wanted to eat rich foods, not salads. You have to want to lose weight to succeed. If you do want to shed some pounds, you know what to do. Cut down on fats and sweets; get more exercise; and, unless you've got a medical condition, the weight will come off.

November 3 - Acknowledging When Great Folks Do Good

I read a story recently about a precious little girl called Alexandra Scott. She is eight years old and has malignant tumors, an aggressive form of childhood cancer called neuroblastoma that starts in certain of her nerve cells. She started selling lemonade when she was four, outside her home in Philadelphia: Alex raised $2,000 in a single day. Each year more and more stands opened all across the country, by people wanting to help out. They are all called "Alex's Lemonade Stand."

This super brave little girl is a role model to us all. She is exhausted and weak from her radiation and chemotherapy, yet she determines to raise money for a cure to be found so that other children don't have to suffer like she has.

What a shining example of everything good in our world. There is such a lot to be done for many, many causes; won't you be an Alex too?

November 4 - Bean Bliss

Sassy Soy bean Casserole

½	cup dried soy beans, soaked for 8 hours
1	tablespoon sunflower or canola oil
1	teaspoon chili powder
1	tablespoon fresh ginger root, grated
1	clove garlic, crushed
3	scallions, chopped
8	ounces mushrooms, sliced
4	stalks celery, cut into julienne strips
2	carrots, cut into julienne strips
4	teaspoons cornstarch
1	tablespoon sherry
1	tablespoon soy sauce or tamari
1 ½	cups soy bean stock
	salt and freshly ground black pepper, to taste

Drain the beans and rinse. Cover with lots of fresh water in a large pan, bring to a boil, and simmer for 2 hours until tender. Drain, reserving 1 ½ cups of stock. Heat the oil in the same pan and sauté the chili powder, ginger, and garlic for 2-3 minutes. Stir in the scallions, mushrooms, celery, and carrots, and cook over a gentle heat for 10 minutes. In a small bowl, blend together the cornstarch, sherry, soy sauce, and soy bean stock—no lumps please! Stir into the vegetables and add the beans. Bring to a boil, and then simmer, covered, for 10 minutes until heated through, thickened slightly, and smelling good! Season to taste and serve hot with brown rice. Serves 6.

November 5 - A Healthy Heart is a Happy Heart

Let's get to the heart of the matter . . . nearly a quarter of a million women die of a heart attack in the U.S. every year; that's more than die of cancer. Heart attacks can be intense and come on suddenly without you realizing, but sometimes your body will give you little warning signs—and they are not all about intense chest pains. Fatigue, shortness of breath, a feeling you have to sit down as you feel nauseous and break out in a cold sweat, you feel anxious or have indigestion symptoms, or you're not sleeping well—these can all be signs of heart attack.

The good news is that everyone is in the position to greatly reduce his or her chance of having a heart attack. You don't have to change overnight, but every one of us owes it to our family not to have them visit us in the emergency room because we were too naive to think it won't happen to us. The sooner you are aware of the risks, and the sooner you begin to implement changes to your diet and physical activity, the sooner you are on your way to a healthier body.

November 6 - Start Now on the Path to Better Health

Diet: Saturated fats found in meat, poultry, dairy foods, egg yolks, shrimp, and scallops all clog arteries with their deadly

Be the Change You Want to See in the World

cholesterol. Vegan foods contain hardly any cholesterol. A bit drastic? Not nearly as drastic as a triple-bypass surgery, or being short of breath, feeling ill, and being on medication for the rest of your short life.

Smoking: Between 20-40 percent of deaths from heart disease are directly related to tobacco smoking. If you quit now it will reduce your chance of a heart attack by half, the benefits will start immediately, and after 5-10 years the benefits are about the same as those who never smoked. Your smoking has to be zero cigarettes though, cutting down, or smoking low-nicotine will not reduce those odds.

Exercise: Being overweight is hard on your heart. Regular aerobic exercise will reduce your risk of a heart attack by up to 50 percent. Again the results can start immediately since your heart is a muscle and exercising to the point of sweating strengthens that muscle.

Stress: Raging on the inside and smiling on the outside is not fooling your heart. People who are angry, frustrated, hostile, impatient, and negative are all personalities prone to heart disease. Chill out, people! Stress kills, and if this is your personality go to stress management classes and change your lifestyle. Women need to be especially careful since many doctors don't screen women for heart attack risk and may even overlook warning signs as hormonal changes or stress. Be warned, two-thirds of women who die of a heart attack had no prior recognizable symptoms. Demand a regular heart screening.

NOV

High blood pressure: Another major contributor to the risk of heart attack is high blood pressure. Cut out animal protein, alcohol, coffee, and salt, lose excess weight, and keep your stress under control.These will all help reduce your blood pressure naturally.

Check out *www.womenheart.org*, American Heart Association, any heart Web site, or vegan Web site for more information. I don't want your name added to the list of people who died of the number one killer in the U.S.; please choose good health.

November 7 - Sleep Your Way to Wellness

Sleep is a very important part of your body's health, along with good nutrition and exercise. Experts tell us we need at least eight hours of sleep a night, but the average woman gets six. Women tend to be the caregivers and put themselves after everybody else, so early mornings and late nights are when we find time for ourselves.

Or maybe you suffer from insomnia—being unable to let the worries of this day and the next leave your thoughts, and so you toss and turn, and are still awake more than thirty minutes after your head touched the pillow. Then you don't sleep right through and wake—thanks to fluctuating hormones, pregnancy, hot flashes, pain, a noisy bed partner, anxiety, feeding a newborn, sinus problems, an uncomfortable bed, or you need to pee. If this lasts for more than a few nights you really must do something about it or your life, and your health, could suffer. If it lasts more than a few weeks consult your doctor, or a sleep specialist.

November 8 - Sleepy Time

Try these suggestions to help you fall asleep . . . and stay asleep:

- Avoid alcohol, coffee, caffeine drinks, chocolate, or smoking close to bedtime. Alcohol and coffee are diuretics and will make you need to pee; the others are stimulants. If you need a snack before bed choose a banana an hour before turning in for the night as it contains an amino acid that is converted to the brain chemical serotonin, which helps regulate sleep.

- Exercise no less than three hours before bedtime to allow your metabolism to slow down. Exercise is good to release the day's stress however.

- Have a nice hot bath or shower before bed to raise your body temperature and destress. Choose soothing lavender bath oils, dim the lights, or use candles to calm the body for sleep.

- Make sure your bedroom is not too cold or too warm, and that it's quiet and dark.

November 9 Take a Nap

If you have a few sleepless nights allow yourself to take a twenty-minute nap between 2:00-4:00 pm to help you through the rest of the day. And never underestimate how important a good night's sleep is; don't let it be a distant dream.

NOV

November 10 - Where Did Your Cup of Coffee Come From?

Listen up, coffee drinkers. Coffee is one of the largest U.S. imports and the second most valuable commodity in world trade after oil. Yet oil barons are mega wealthy, whereas many small coffee farmers receive prices for their crop that are less than their production and living costs. This leaves them in constant debt and poverty.

So who gets all the money? The huge plantation owners, and the middleman to whom the coffee farmers sell their crop.

In the 1980s, Mexican coffee farmers let the U.S. know they would rather have a fair price for their coffee than accept charity aid, "If we are going to get anywhere we (the coffee farmers) must have access to real markets." And so the Fair Trade Network was established to assure the farmers of a fair price. Now many coffee farmers have a living wage, can stay on their farms, plan their futures, feed, clothe, and educate their children . . . and they have regained their dignity.

Fair Trade coffee now supports small farmers throughout the coffee-growing world. Not only are their wages improved but small farmers do not cut down huge swathes of rainforest or buy huge amounts of chemical fertilizers and pesticides. Many practice and pass down to their children a sustainable, organic farming technique, which leaves the soil healthy, uses no pesticides to harm their children, the environment, birds and wildlife, or their water supply.

We too can support their endeavor: coffee grows well in the shade, so look out for and buy "Shade Grown, Fair Trade certified coffee." Yes, it's more expensive, but by paying about 3 cents more per cup you are directly benefiting the small farmer producer. In coffee shops too ask for Fair Trade coffee each time. Ask for it at the market.

November 11 - Get the Great Skin Glow

Your skin is the largest organ of your body and is of vital importance in eliminating wastes from your body. Dry skin brushing is a great way to exfoliate your skin by sloughing off dead skin cells. It clears pores and encourages cell renewal. It stimulates the body's circulation, strengthens the immune system, aids digestion, helps disperse cellulite, and takes some of the burden off the lymphatic system that transports cellular wastes, bacteria, and viruses. By removing dead surface cells the skin is oxygenated and new cell renewal is increased, helping to prevent premature aging and degenerative diseases. Your whole body glows with better health.

To dry skin brush you'll need a natural paddle brush made of natural plant fibers that doesn't scratch the skin. Mine is cactus spikes, but you can also buy sisal from your beauty counter. Don't choose plastic, nylon, or boar bristles; these are too stiff. You may also want to choose one with a removable long handle to reach all parts of your body.

• The massage should only take 5-10 minutes a day. Always brush on dry skin to remove dead skin cells and to avoid stretching the skin.

• Start with the soles of your feet and use circular motions. Then brush up your ankle, calf, then thigh and repeat with the other leg. Don't overdo it and don't press too hard. Brush each area several times and always towards the heart, going with the body's own blood and lymph circulation.

- Do light, circular strokes over your abdomen and over and around your breasts, avoiding your nipples. Repeat the circular strokes on your buttocks, and I do extra on my outer thighs for that awful cellulite.

- Brush your back and across your shoulders.

- Brush up your hands and up each arm, front and back, towards the armpits and shoulders.

- If you want to you can brush your scalp, too.

- Avoid any irritated areas, or cuts or burns.

- I like to follow with a shower to remove the loosened skin.

- Store your brush in a dry place, and wash it every few weeks to prevent bacteria forming.

November 12 - Nuts About Nuts

Nuts are one of the most perfect snacking foods you could choose. Yes, they are fairly high in calories, but they are also very nutritious and are high in protein, minerals, fiber, and unsaturated fats. The American Institute for Cancer Research recommends walnuts as a source of omega-3 fatty acids; a handful a few times a week may well reduce your risk of a heart attack.

November 13 - Rethinking Resources

Do you ever wonder about the direction our planet is advancing—if, indeed, it is advancing? Did you know the Earth's population has doubled since 1950 and we are consuming its resources 20 percent faster than Mother Nature can replenish them? In the last fifty years oil consumption has increased sevenfold and in the last one hundred years the use of fresh water has increased sixfold. The gap between rich and poor is ever increasing with over one third of the population of our planet suffering malnutrition. What on Earth are we doing? We have no right to behave so irresponsibly: our Western selfish attitude will surely leave our todays with no tomorrows if we don't significantly change our ways.

November 14 - Meat Industry and Disease

More concerned about Mad Cow Disease now that it has hit our shores? Just what is it and where does it come from? In sheep it's called "scrapie"; in cattle, Bovine Spongiform Encephalopathy, or BSE; in humans it is called, "Creutzfeldt Jakob Disease" (CJD). All these manifestations are a progressive nervous system disorder and, at present, there is no cure.

It is transmitted by eating an infected animal, and the epidemic started when infected sheep with scrapie were fed to cattle. Wait a minute, I hear you say, cows are herbivores, they don't eat meat—well, it used to be like that when animals were kept in fields, but with intensive farming it's common practice to feed dead animals to cows to increase milk yield, etc. Since the disease is not killed by cooking, sterilization, pasteurization, freezing, or any other food preparation technique, the animal, which includes humans, that eats the infected meat may well contract the disease too.

When are people going to realize that you cannot mess around with nature in this way, and expect the world to continue safely and for the sole gain of the human race? How many different ways are there to say it? Eating a whole foods, vegan diet is the only truly healthful and sustainable choice.

November 15 - Dining on Unhappiness

Why are factory-farmed animals fed so many antibiotics in the first place? Because thousands of pigs, cows, or chickens live in one huge smelly shed, and the drugs are routinely added to the animals' feed to prevent infection before, and in case, it happens. Also added to their feed are hormones and synthetic vitamins for growth stimulants, and tranquilizers to keep the animals calm during their terrifying ordeal of just being alive.

All these additives are ingested by the animal when they eat, used in their bodies, stored in their flesh, and some are elim-inated in their natural waste products, but when the animal is slaughtered those drugs stop being eliminated—when you eat a steak or a chicken breast, you also are eating all those additives stored in that flesh. As soon as the animal is dead it begins to decay, bacteria start to multiply—like your steak rare?—not all the bacteria is killed in cooking, and you get food poisoning. You can't see salmonella with the naked eye, but salmonella can kill, and it's on the knife, the counter, and the chopping board. In humans it causes vomiting, and diarrhea; the elderly and young children are particularly at risk—that 24 hour flu could well be food poisoning.

Antibiotics are a wonder drug of our lifetime, they have saved millions of lives, but 60-80 percent of factory farm livestock are now regularly given the same antibiotics in this country. The DNA of the bacteria these antibiotics have historically eliminated are changing, making the disease now resistant to the drug. These bacteria, such as E. coli and salmonella, can now survive in the animals; when you eat meat that is contaminated and your doctor gives you traditional antibiotics to help, this now leaves you vulnerable to life-threatening diseases. In 1997 the World Health Organization called for a ban on using human antibiotics just as a matter of course in animal feed. Don't risk it. Go vegan.

November 16 - Giving the Gift of a Better Life

Looking for an unusual holiday gift? Click onto *www.heifer.org*. They claim to have the most important gift catalogue in the world. On this site you can "buy" a llama for $150, a goat for $120, a heifer for $500, or bees, a pig, buffalo, rabbits, ducks, or geese. No, not for a farm of your own. As a donation for a person in need in another country.

Heifer says, "Change comes slowly" but with more and more people like you and I, these poorer families can eat, live, and even dream of their children's future. I know I am vegan, but I am able to go to my local store and choose soymilk and a vast array of various vegan proteins and choices of fruits and vegetables. Besides, these animals will be well cared for, fed natural foods, and even cherished, much as my dog is, they will not be kept in abject misery and total cruelty on factory farms like something that doesn't matter.

November 17 - Gifts That Keep on Giving

Buying an animal that will save other children from starving to death is a great way to make a difference, don't you think?

November 18 - Fur is Unfashionable

From November 15 to May 15, Canada's commercial harp seal hunting season is open. The main culling takes place in the springtime when the young seals are being born—indeed almost 95 percent of the seals killed are between 12 days and one year old—but it's not too early in the season to become aware of this barbaric practice. The estimated total kill of Northwest Atlantic harp seals for 2002 was over 580,000 animals. The younger seals are clubbed to death on the ice, or killed with hakapiks, which are like heavy ice picks. The older seals are shot with a rifle, both on the ice and when swimming in the water. Nearly half the baby seals have to be struck a second time because the first blow doesn't kill them, and many seals were still alive when hunters skin them.

Why are these seals hunted like this, and in such large numbers? The International Fund for Animal Welfare (IFAW) says, "A cull is designed to reduce the size of a population; in this case, ostensibly to reduce the impact on fish stocks, and provide benefits for commercial fisheries."

I cannot believe that anybody could look down into the limpid eyes of a two week old, fluffy white baby seal, lying on the ice, raise a club and smash it down on that beautiful head, then raise it again and again until the virgin snow is saturated in red. How far has the human race come that we can still be so unashamedly cruel and wicked? We must not sit back and allow things like this to happen, just to turn our heads so that people can eat the fish that is the seals'. Please go to www.canadasealhunt.ca to find out what you can do.

November 19 - We CAN Feed the World

The fact is that there is enough food in the world for everyone. But tragically, much of the world's food and land resources are tied up in producing beef and other livestock—food for the well off—while millions of children and adults suffer from malnutrition and starvation.

—Dr. Walden Bello

November 20 - World Children's Day

Today is World Children's Day: More than 30,000 children under the age of five die, *every day of the year* throughout our world, of hunger and other preventable diseases.

Organize fundraisers to sponsor a child, have local toy drives for children in hospitals, and keep your eyes and ears open for other ways to help. We all live in this world together, there is no them and us, we are all children of our planet.

Check out the Web sites of Save the Children, Unicef, Global Movement for Children, ActionAid, Plan International, Child Reach, Children's Defense Fund, and others, for more information.

November 21 - Furry Friendships

Holiday times should be happy for humans, yes, but also healthy times for our companion animals. The food, decorations, and entertaining we take for granted need special consideration for our furry friends.

* Dogs and cats love routine and can be very distressed when their lives are changed by your entertaining family and friends, extra shopping, time spent in the kitchen, and your children's extra activities. Be sure your dog or cat isn't left out: Fido still needs his long walks, and Fluffy her playtime. During Thanksgiving be sure you let them know you are thankful for their love, loyalty, and the joy they bring you.

* Holiday food is too rich for us, never mind your canine and feline companions so don't even think about overindulging them unless you want to clean up puke and upset tummies from your carpet. Bones are very dangerous for dogs and cats as they splinter and can become lodged in their throat and intestines. Don't leave party food unattended in the kitchen, or the coffee table—far too irresistible to Fido.

* Watch the decorations—gourds, pumpkins, and corn can be tempting chew toys, but they are dangerous and can jam up the digestive system when swallowed. Holly and other plants brought indoors are not a salad bar and can be toxic.

* Ribbons, foil wrapping paper, and tinsel are very pretty and can be fun for Fluffy to play with, but again, could get lodged in her throat and digestion. Clean these up immediately.

Be the Change You Want to See in the World

- Extra guests mean more doors being opened and less super vision for Fluffy and Fido, so place them in a quiet, closed room for extra nap time. Give them a little treat and their favorite toys, and check periodically. Tell guests to respect your animals and leave them to snooze.

- Be careful of lit candles—very tempting for Fluffy to investigate.

- Keep the phone number of an emergency vet on hand . . . just in case.

- Never, ever give any animal as a gift. Instead wrap a food bowl or a toy with a tag promising to help the recipient find a homeless furry friend after the holidays.

- Holiday pet perils can be happily avoided with common sense, awareness, and love. Make sure your companion remains safe and healthy. Visit *www.aspca.org* for more detailed information.

November 22 - Thinking About Thankfulness

This is certainly not a happy time of year for turkeys. To satisfy our nation's nostalgic yearnings, tens of thousands of turkeys live crammed into warehouses where many will be smothered alive, be riddled with disease, or suffer from heart attacks. When they have grown grossly and unnaturally huge due to feed full of hormones and antibiotics the turkeys will see the light of day—they are shoved into trucks for the trip to the slaughterhouse. Here their throats are slit while fully conscious, or those who missed the automated knife will be scalded alive. Are these practices you really want to support?

Is it really necessary to include a turkey in our happy celebration of life, the bounty of the harvest and things we are thankful for?

November 23 - Veggie Thanksgiving Options

There are so many wonderful vegan dishes for people to enjoy with incredible flavors, colors, and textures in the grains, and fruits and vegetables of the season. If your family really needs "meat," choose a Tofurky, a stuffed soy alternative, which you can buy frozen along with wild rice stuffing and gravy.

The following days contain some of my favorite recipes for an animal-friendly Thanksgiving celebration:

Lemony Vegetable Parcels

2	medium carrots, scrubbed and cubed
1	small rutabaga, peeled and cubed
1	large parsnip, peeled and cubed
1	leek, washed well and sliced
	finely grated zest 1 lemon
1	tablespoon lemon juice
1	tablespoon whole grain mustard
	salt and freshly ground black pepper, to taste
	olive oil

Preheat oven to 375°F. Place the carrots, rutabaga, parsnip, leak, lemon zest, lemon juice, and mustard in a large bowl. Toss lightly to mix well. Season to taste. Cut four 12-inch squares of nonstick cooking parchment—waxed paper is fine, but foil will not work.

Brush the squares lightly with the oil. Divide the lemon veggies between them. Fold over the tops and tuck the sides underneath to form a package. Be sure they are sealed well. Transfer these parcels to a baking sheet and bake for about an hour or until just tender. Serve on plates as a side dish to the vegan roast. Let guests open each parcel to reveal the contents. Mind the steam! Serves 4.

November 24 - Bowls of Abundance

Colorful Harvest Gratin

2	large sweet potatoes, peeled and sliced ¼ inch thick
2	medium carrots, sliced on the diagonal
1	small rutabaga, peeled and cut into ¼ inch chunks
2	tablespoons olive oil
1	medium onion, chopped
1	clove garlic, crushed
1	inch fresh ginger, grated
1	teaspoon turmeric
1	teaspoon ground coriander
3	teaspoons ground cumin
2	teaspoons lemon juice
½	cup sunflower seeds
1	cup grated vegan cheddar cheese

Preheat the oven to 375°F. Bring a large pot of water to a boil and add the sweet potatoes, carrots, and rutabaga. Boil for 5 minutes. Drain, but reserve ½ cup of the cooking liquid. Tip these veggies into an oiled 9x13 ovenproof dish. In the original pan, heat the oil and sauté the onion, garlic, ginger, and turmeric for 5 minutes, stirring often. Stir in the coriander and cumin, and then add the sweet potato mixture. Toss gently and heat through for a further 5 minutes. Stir in the reserved ½ cup cooking liquid and the lemon juice. Check for seasoning before tipping into the baking dish. Cover and bake for an hour. Remove the cover and sprinkle with sunflower seeds and vegan cheddar. Return to the oven for 10 minutes, then allow to cool before serving. Serves 4-6.

November 25 - Yummy and Unusual Treats
Holiday Nut Roast

1	cup fresh whole wheat breadcrumbs
½	cup cashew nuts, chopped
½	cup hazelnuts, chopped
½	cup sunflower seeds
1	apple, peeled and grated
1	carrot, grated
½	cup fresh herbs—such as thyme, rosemary, parsley, or oregano—chopped
	salt and freshly ground black pepper, to taste
1-2	teaspoons olive oil
1	medium onion, chopped
1	clove garlic, crushed
4	ounces mushrooms, sliced or chopped
½	red bell pepper
½	green bell pepper
1	teaspoon soy sauce or tamari
1	cup tomato juice

Preheat oven to 375°F and oil a loaf pan. In a large bowl, combine the breadcrumbs, nuts, sunflower seeds, apple, carrot, herbs, salt, and pepper. Heat the oil in a large skillet, sauté the onion, garlic, mushrooms, and peppers for 5 minutes or until softened and a nice color. Season again and then stir into the breadcrumb mixture. Stir the soy sauce into the tomato juice and then add to the rest of the ingredients, mixing enough to form a dropping consistency. Tip the mixture into the oiled pan, smooth the top, and bake for about an hour until golden brown and cooked through. Serves 4-6.

November 26 - Bake an Organic Dessert
Cranberry and Apple Tart

Crust

1½   cups whole wheat flour
      pinch salt
1½   tablespoons brown sugar
6   tablespoons vegan margarine, chilled and diced
3   tablespoons ice cold water (or more if needed to amalgamate the flour to dough)

Filling

4   medium Granny Smith apples, peeled and sliced ¼ inch thick
¾   cup fresh cranberries, rinsed
3   tablespoons whole wheat flour
⅔   cup granulated sugar
1   teaspoon ground cinnamon

Topping

1/2   cup rolled oats

6   tablespoons whole wheat flour

1/3   cup packed brown sugar

3   tablespoons vegan margarine, melted

In a food processor, combine 1½ cups flour, pinch salt, and 1½ table-spoons brown sugar. Whiz for 30 seconds. With machine running, add 6 tablespoons margarine through the feed tube, and then add the water. Whiz some more until a ball forms. Remove and wrap dough in cling wrap, place in fridge, and chill for at least 30 minutes. In a medium bowl, gently mix the cranberries, 3 tablespoons flour, 2/3 cup granulated sugar, and cinnamon and set aside. In another medium bowl, combine the oats, 6 tablespoons flour, and 1/3 cup packed brown sugar. Stir in the 3 tablespoons of melted margarine. Lightly butter a 10-inch tart pan with a removable base. Roll out the chilled pastry and fit it into the bottom and sides of the pan, cover, and refrigerate until ready to use. Preheat oven to 375°F. Spoon the fruit filling into the pastry base. With your fingers, crumble the topping over the fruit. Bake for 40 minutes. Allow to cool for 10 minutes and serve decadently with soy ice cream. Serves 6-8.

November 27 - Turn off Your TV and Turn Your Brain Back on

Nowadays children have computers and chat rooms, mobile phones, and a TV in their room with a hundred channels in glorious color, stereo, and DVD. Children are techno-wizards. TV is a time filler, and time killer. It brainwashes. It both encourages and plays down violence, in both behavior and attitude. Every ten minutes is a fast food ad, and ads for wanna have, gotta have. TV makes children lazy, sluggish in thought and action.

- If you feel your children are watching too much television and it's having a negative impact:

- Have cable disconnected.

- Remove all TVs except one from the house.

- Limit TV viewing to set hours, such as only after homework is finished, no TV during meals, no morning TV.

- Limit channels watched.

- Have family evenings that are fun. Encourage new hobbies.

- Don't rely on the TV for entertainment; make your own. Talk to your family, and have them talk to you.

November 28 - Pet Seat Safety for Cars

Never let your dog climb all over you as you drive, sit in your lap, or in any other way distract you in the car. It's not good for you or the dog. Dogs can be thrown through windshields in an accident too. You can buy safety belts specially made for you to buckle your dog into, or train her to stay in a fastened crate on the back seat or folded down seats.

November 29 - Soulful Soul Food

Now that winter is coming on, it's time to think about hearty, soulful, warming foods. What better than beans? Beans are edible seeds that are formed in a pod, and can then be eaten fresh, dried, sprouted, or ground to make flour. They include peas and lentils, and alfalfa. This food group is extremely nutritious, being high in protein and carbohydrates, good sources of iron, zinc, and calcium, vitamins A, B1 and B2, and a great source of fiber. They also lower cholesterol, blood sugar, and are thought to lower risk of cancer, especially in the colon.

Canned beans are excellent to use; just be sure to rinse them well. Dried beans should be soaked for about eight hours in plenty of filtered water, drained, rinsed again and then gently simmered in four times as much fresh water as beans. Red kidney beans should be boiled for ten minutes to kill any harmful enzymes. Never add salt at any stage as this will toughen the skins and make them more difficult to digest. Cook the beans for 1-2 hours depending on the type (lentils and many other legumes cook even faster) and rinse well again to use in your recipe—this makes it easier on your digestion and helps eliminate gas. Follow the directions on the packet for cooking times. Or use the quick soak method: boil rinsed beans for 5-10 minutes, turn off the heat and allow to soak for 2-3 hours. Drain and rinse, then cover with fresh water and cook according to directions.

November 30 - Beautiful Beans

Beans are extremely versatile and can be used in soups, salads, and main courses—try them all!

Thousand Bean Soup

2	cups of mixed dried beans and lentils
10	cups vegetable stock
1	onion, chopped
2	cloves garlic, crushed
1	medium potato, scrubbed and diced
2	stalks celery, chopped
2	carrots, scrubbed and chopped
3	bay leaves
1	teaspoon dried oregano
	sprigs of fresh thyme
1-2	tablespoons tomato paste, to taste
	salt and freshly ground black pepper, to taste

Soak the bean mix for 8 hours in cold water. Drain and rinse. Bring plenty of fresh cold water to a boil in a large pot. Boil the beans for 15 minutes, drain, and rinse. Return the beans to the pot and stir in the stock, onion, garlic, potato, celery, carrots, bay leaves, oregano, and thyme. Bring to a boil, lower heat, and simmer for 1 ½ hours. Remove bay leaves. Cool slightly in batches, pour into a blender, and blend until smooth. Return to the pot, reheat, add tomato paste, and season with salt and pepper. Check the seasoning and add more fresh herbs if the soup needs more flavor. Good the next day too. Serves 6-8.

DECEMBER

Peace and Reflection

December 1 - Uniting to Prevent and Cure AIDS

It's World AIDS Day around the globe today. Unfortunately, the breakthrough medicines we have access to in this country are not available in the quantities necessary to save people in Third World countries, and along with poor nutrition, millions are still dying. It is tragic.

HIV is transmitted from person to person in the exchange of body fluids, such as semen, vaginal secretions, and blood, through sexual intercourse, infected blood transfusions, or sharing drug needles.

Educate yourself about HIV and AIDS; they are very serious, potentially deadly diseases, and it is up to each one of us not to spread them. Get tested for HIV yourself and, if you test positive, get the care and support you need. If you have children be sure they are fully educated on the terrible dangers of HIV and unprotected sex; talk to them openly and don't shrug off any embarrassment you or they may have. The price is too high.

December 2 - Do What You Can Do To Help

We worry so much about terrorism in this country, yet AIDS is the most intimidating terrorist in the world. According to the Web site *www.apathyislethal.org* AIDS has currently orphaned 14 million children . . . and could reach 40 million within the next few years. Numerous charities and agencies are working hard every day by going into the poorer countries to educate and break the silence that surrounds this disease. It is the ignorance, shame, and fear that must be defeated by changing people's attitudes.

UNICEF advocates the increased access of life-saving information, particularly to young people, as the most important measure against the risk of getting infected. The proportion of infected girls to boys is 2:1 in sub-Saharan Africa, where 77 percent of young people infected with HIV live. It is the girls and women who are more vulnerable because their society expects them to be ignorant about sex. They also face a stigma if they use family planning; often it is their husbands who will not allow them to do so. Females are also used in the sex trade for money, food, and shelter. In parts of Africa it is believed that sex with a young virgin girl will cleanse a man of his HIV virus; the girls do not know that someone with the HIV/AIDS virus can still look healthy.

It is all about education. By changing attitudes there is a chance that the current course of the spread can be reversed, and UNICEF believes that young people hold the key to stopping the AIDS epidemic: they are the ones more likely to adopt, and maintain, safe behaviors.

We can help by supporting and donating to AIDS organizations. I read on Apathy is Lethal's Web site that the Salvation Army's Masiye camp in Zimbabwe teaches AIDS-orphaned children life skills and promotes psychosocial support. Since it started in 1998, it has treated 3,000 children.

This AIDS epidemic is an unparalleled challenge to humanity; let's shake off our apathy and do what we can, in every small way, to show we care for these people and their children: we are all in the human race together.

December 3 - I Am the Change, You Are the Change

When I was young and free and my imagination had no limits, I dreamed of changing the world. As I grew older and wiser, I discovered the world would not change, so I shortened my sights somewhat and decided to change only my country. But it, too, seemed immovable.

As I grew into my twilight years, in one last desperate attempt, I settled for changing only my family, those closest to me, but alas, they would have none of it.

—**From the tomb of an Anglican Bishop**

December 4 - Keep the Cards and Letters Coming

Have you bought or written your holiday cards yet? If not, this year think about buying cards that support a charity, such as UNICEF, the ASPCA or other humane societies, a research organization, or a local organization. You can pick your favorite. When you have your cards, allow yourself a couple of hours by the fireside to write and address them. Don't think of it as a chore, but as a loving reminder of the people you know, people who have touched you, and who are a part of you. This is a good time to let them know you are thinking of them. If you can possibly swing it, hand address the cards. And sign your name to each, including a few words of fondness and concern. Our lives are so rush, rush, busy, busy; this year make your holidays cards a special effort for remembrance and love.

December 5 - Energy Pick Me Ups

At this time of year you may not be eating a lot of salads, preferring those hearty stews and veggie casseroles. But please don't forget sprouted seeds and legumes, which make an instant side salad with a squeeze of lemon and a little seasoning. I also love them in a lunch sandwich with peanut butter, baby spinach or spicy arugula leaves, and tomato and avocado slices. Sprouted legumes are very nutritious, being high in vitamin C, protein, minerals such as calcium and iron, and enzymes, and after five or six days germination also contain vitamins E, K, and others. You can buy ready sprouted seeds and beans in your whole food market, such as alfalfa, mung bean sprouts, lentil, wheat, rye, clover, or broccoli sprouts, or it's easy to grow your own. Your local health food store should have special sprouting equipment and the grains and beans you should sprout. Growing sprouts can be a kid-friendly, fun project too.

December 6 - Veggie Fruitcake

Ah, ubiquitous fruitcake, your season has arrived! You either love fruitcake . . . or you hate it! I love it—rich, dark, moist with a hint of brandy.

Of course the recipes for mince pies, puddings, and cakes passed down from my mother contained eggs, lard, and butter. Here's a wonderful recipe that has become my favorite. It's adapted from John Robbins' "Ginger Carrot Cake with Orange Glaze" from his book, *May All Be Fed: Diet for a New World*, which has many great vegan recipes, as well as educated reasons for becoming vegan. For this time of year I like to substitute some of the raisins with dried cranberries. This cake is wonderfully moist and keeps well in the refrigerator for up to a week . . . not that mine ever last that long—they are usually eaten within two days! This cake is also good baked in a bundt tin, wrapped in bright cellophane, and makes a very acceptable hostess gift. Happy Yule.

Organic Carrot Cake

3	medium organic carrots, peeled
½	inch fresh ginger, peeled
1½	cups dried apricots, cranberries, or raisins
2	cups whole wheat pastry flour
1	teaspoon baking powder
1	teaspoon baking soda
1	teaspoon ground cinnamon
1	teaspoon freshly grated nutmeg
¾	cup orange juice
½	cup canola or sunflower oil
½	cup pure maple syrup
	a few drops of orange oil
1	cup chopped walnuts or pecans, plus a dozen halves to decorate the top

Orange Glaze

2	tablespoons maple syrup whisked with 1 tablespoon orange juice

Preheat oven to 350°F. Lightly oil an 8x8 inch cake pan, or use a bundt pan if you have one. In a food processor, whiz the carrots and ginger to grate, and place in a small bowl. Do the same to the dried fruit. In a large bowl, whisk together the flour, baking powder and soda, cinnamon, and nutmeg. Then put the orange juice, oil, maple syrup, orange oil, plus ½ cup dried fruit in the processor and whiz until smooth. Add the carrots and ginger, and pulse just to mix. Pour this into the flour mixture and combine, using as few strokes as possible, until the flour is incorporated. Do not over mix the batter. Fold in the remaining cup of dried fruit and the chopped nuts. Tip into the prepared pan and spread evenly. Decorate the top with the nut halves. Bake for 35-45 minutes (40 minutes for the bundt pan), until an inserted toothpick comes out clean. Let the cake cool in the pan for 10 minutes then transfer to a wire rack.

While the cake is still warm, drizzle the glaze carefully over the cake and allow to cool before slicing. Serves 9.

December 7 - Enjoy the Simple Pleasure of Life

Before things get too hectic, plan some time with your friends . . . how about a poetry evening, or reading of a favorite piece from a book? Ask friends to bring along their favorite poem to read and share. Serve some simple, healthful goodies to eat and drink, decorate the room, light the fire, pull up some extra chairs, and enjoy the company!

• Warm some apple cider in a large slow cooker. Add an orange or lemon studded with cloves, a couple of cinnamon sticks, and a diced apple. Place holiday paper cups and a ladle next to it for guests to help themselves.

• Brew a spicy blend of decaf coffee or tea.

• Make a large pot of your favorite soup or a homemade veggie chili and leave simmering. Serve either with warm rolls.

• Add sliced cake, or cookies to end on something sweet.

December 8 - Quality, Not Quantity

Hate this time of year because you know you will put on pounds, and then have to struggle in the New Year to lose them again? You can solve that problem by not overeating in the first place! I'm not telling you not to enjoy yourself, just cut down on the quantities you eat and drink: have a few chips, peanuts, olives, and other high-fat foods, not a whole bowl-ful. Have one glass of wine, not a whole bottle, and one or two cookies, not the whole plate.

As always, walking is the best way to keep the weight off . . . walking away from the buffet table! And if you are sitting at a table with nuts and potato chips, move them down to the other end. If you have a party, ask guests to take leftover food home with them, or dump it in the trash rather than stashing it in your fridge. If friends or work colleagues give you presents of cookies, chocolates, or those huge cans of popcorn, don't open them, but pass on to a children's home, or a skinny relative!

Never grocery shop when you are hungry or you'll stock up on pastries, snacks, cheeses, and other calorie-high yummies. When you eat out, avoid dressing on the salad, avoid desserts and the cheese tray, and ask for a spoonful of your partner's instead . . . just a spoonful, mind you, not the whole thing!

December 9 - Joining Together to End Violence

Today is Human Rights' Day. We tend to turn our heads and pretend all's right with the world, but it isn't. People in other countries, especially women, do not have the luxury of being

Be the Change You Want to See in the World

in the "land of the free." People all over our world are suffering from torture, imprisonment, and execution unjustly and unfairly, and it must stop. It is up to each one of us to educate ourselves and express our concerns.

Log on to Amnesty International's Web site, *www.amnesty.org*, and read about the discrimination and violence against women, about rape in Kenya and Mexico; violence against women in Pakistan, or the fact that Saudi Arabia has one of the highest rates of execution in the world, many for nonviolent crimes without a fair, public, and represented trial. Bookmark Amnesty International's Web site and check back to it weekly; help them protect human rights around our world. They really are making a difference, and they are being heard. Make yourself and your voice heard too.

December 10 - Diet for a New Year

You don't have to munch on just carrot and celery sticks all season—eat and drink sensibly, choose a small plate at the party buffet, and don't visit again and again. Then, on New Year's Eve you can wear that same little dress that saw in the new millennium with you.

December 11 - Compassion is High Fashion

Fur used to turn heads, now it turns stomachs.

—Rue McClanahan

Behind every beautiful fur, there is a story. It is a bloody, barbaric story.

—Mary Tyler Moore

We don't live the lives of Eskimos. We don't need to kill animals for fashion.

—Charlize Theron

December 12 - No More Fur

When buying your gifts this year, be sure there are no fur trims attached. And be sure your family and friends know where you stand on fur items too. Don't support department stores that sell fur coats, and let them know why you no longer shop there. Support everyone who is drawing attention to the cruelty of fur. Print out little quotes, like the ones from the three ladies above, and when you see people wearing a fur coat, stole, or hat, give them your printed paper and say, "Oh, you dropped this!" and watch, or not, as they read it. It's nonconfrontational yet the message gets across; maybe they'll think twice next time they buy a coat. The only place for fur coats is on the wild animals they belong to.

December 13 - Great Friends Are the Best Gifts

The ornaments of a house are the guests who frequent it.

—Anonymous

December 14 - Be the Host with the Most

Who hosts the Yuletide celebrations in your family? If it's the same person every year, make sure she really wants to: talk to her about it in person if possible, her eyes and body language will tell you the truth. Maybe it's time for a change. My mother was perfectly happy handing over the reins to me

Be the Change You Want to See in the World

when my partner and I bought a house with a huge dining/sitting room. I love cooking, and now my mother could relax and enjoy the festivities.

Our food leftovers are put outside the patio doors for the birds and foxes to enjoy—who needs TV; we can watch nature.

December 15 - Pop Some Corn and Play a Game Together

This holiday time, try to turn on your TV as little as possible. Arrange some games that adults can play—board games, charades, and monopoly. If you have talented people in your family prepare them a few days in advance, and get them to sing a song or two, recite a poem, or tell a story; wherever their talent lies. This holiday, don't let TV take over.

December 16 - Small Comforts for Hard Times

If you catch a cold or the flu this winter, have a stock of feel-good foods and medicines on hand, lay back, and watch some old movies. Don't struggle into work and infect everyone else. Be kind to yourself, be indulgent, get your mom to come 'round and take care of you!

December 17 - Brilliant Advice from Mom

If you treat a sick child like an adult and a sick adult like a child, everything usually works out pretty well.

—**Ruth Carlisle**

December 18 - Quiet Your Mind and Heal Your Body

After a long, tiring day at work or the mall, set aside some time for yourself. Pour yourself a glass of wine or sparkling water, put on some relaxing music, light some scented candles, lock the door, and immerse yourself in your stress-free zone.

If you are really wound up, stock up on aromatherapy products with scents of lavender or chamomile to soothe and lessen anxiety. Rose works well too—buy candles to match. Ask your florist to save you her past-their-best roses, and sprinkle the petals on the bath water's surface for extra luxury.

When your stress has dissolved away, wrap yourself in a huge soft towel or robe, and promise yourself this will now be a regular indulgence. You owe it to yourself.

December 19 - Consider a Less "Consumer Christmas"

I don't read Charles Dickens as a rule, but *A Christmas Carol* is a must at this time of year. It is so evocative and thought provoking. It reminds us that this is the season for benevolence. Any Scrooges need to see the ghosts of Christmas past and Christmas future, and change into loving and giving human beings.

Do we really need all these gifts, all this food, and all this stuff? Make this the season that your family gives to those less fortunate.

December 20 - Donating is Great Karma

• Ask your child to give one of their gifts to a less-fortunate child.

• If you host a party ask friends to bring a gift suitable to donate to a children's charity, nursing home, or an animal shelter, instead of a hostess gift or bottle of wine.

• Donate some of the money you would have spent on expensive champagne or liquor to a local soup kitchen or hostel for the homeless.

• Your local newspaper may print more ideas for those in need, please don't ignore their requests. Be generous.

December 21 - When the Sun is Reborn

Yule, also known as the Winter Solstice, occurs when the sun enters Capricorn. It is the shortest day of the year; it falls between the 20th and 23rd of December.

In ancient times, the people believed the sun needed their help, so they lit huge bonfires to strengthen the sun and show it the way back. The solstice was the date when the days started to become longer, and light would return to the land, giving the people cause to celebrate and look forward to the warmer spring.

Whether we celebrate a Pagan Yule, Christmas, Kwanzaa, or Chanukah, all across the world we must stop harboring resentment, hatred, anger, and mistrust. Religions must be about forgiveness, not violence and killing, and certainly not ethnic cleansing. It is a time for peace and goodwill to all men, women, and children.

December 22 - Spices are Medicinal

I love a good, spicy curry to add a different note to my vegetarian cooking. Curries are also warming, and their seasonings help relieve cold and flu symptoms. They are also good to serve to meat eaters, who will eat heartily without even realizing the meal is meat-free.

A variety of vegetables can be used, including potatoes, spinach, cauliflower, carrots, peppers, and celery, and many of the pulses: garbanzos, black-eyed peas, and various lentils and split peas. It's the spices that are the main players—cumin, ginger, coriander, turmeric, fenugreek, chilies, and others. To complete the meal, serve with side dishes of basmati rice, and bread such as naan or chapatti. Personally, I prefer the more delicate curries of Southern India, which include coconut milk in the ingredients. Curries can be as complex, or as simple as you like, and as hot or mild . . . experiment! Wash down with a cold beer and finish with a refreshing sorbet.

Punjabi Potato Curry (Aloo Gobi)

4	tablespoons canola or sunflower oil
1½	pounds new potatoes, peeled and cut into large chunks
1	yellow onion, chopped
3	cloves garlic, crushed
1	teaspoon garam masala
½	teaspoon turmeric
1	teaspoon ground cumin
1	teaspoon ground coriander

Be the Change You Want to See in the World

1   inch piece fresh ginger root, peeled and grated

1   fresh red chili, chopped

1   small cauliflower, separated into florets

4   tomatoes, peeled and quartered

1   cup frozen peas

1¼   cups vegetable stock

½   cup chopped cilantro, plus extra to garnish
     cooked basmati rice and warmed Indian bread

Heat the oil in a large heavy pan. Stir in the potato chunks, onion, and garlic and sauté over a low heat for a few minutes. Stir in the garam masala, turmeric, cumin, coriander, ginger root, and chili, and sauté for another minute, stirring all the time. Add the cauliflower, tomatoes, peas, and stock. Cover and cook over a low heat for 30-40 minutes until the potatoes are tender and cooked through. Stir in the cilantro. Tip into a warmed serving dish and garnish with the cilantro. Serve with the rice and bread and a bowl of lentil dal. Serves 4-6.

December 23 - A December Yummy Yule

Spicy Dal

1   cup dried red lentils (masoor dal)

½   cup dried yellow split peas (chana dal)

4   cups water

1   small onion, finely chopped

½   teaspoon turmeric

1   cup fresh tomatoes, chopped

½   teaspoon ground coriander

2	teaspoons ground cumin
2 ½	cups water
½	cup fresh cilantro, chopped

In a large pot, bring the lentils, split peas, 4 cups water, onion, and turmeric to a boil. Lower the heat and simmer for 30 minutes, stirring occasionally, until the lentils are tender. Stir in the tomatoes, coriander, cumin, and 2 ½ cups water. Bring to a boil, lower heat, and simmer for a further 15 minutes, covered. Stir in the cilantro at the last minute and serve. Serves 4-6.

December 24 - Peace and Goodwill to All

When the relatives start arriving, take a deep breath, give sincere hugs and smiles, and see the best in everyone. Ban gossip and bad mouthing right from the start; every one of us has faults, let's not air them. Either change the subject, give the person a task to do, or simply tell them, plainly and directly, that this is the season of goodwill.

Alcohol can make matters worse, so offer nonalcoholic punch, make weak cocktails, serve wine in small glasses, and make tea and coffee instead of serving brandy. We've all seen people get nasty when they've had too much to drink, and besides, you want your guests alert to help you with the clearing up.

December 25 - Have a Gentle Heart

You cannot shake hands with a clenched fist.

—**Indira Gandhi**

December 26 - Try a Healthy Detox Fast

So, how has the season left you feeling? Jaded and listless, or happy and uplifted? If you are feeling run down and sluggish, try some natural remedies rather than resorting to over-the-counter medicines full of additives and chemicals.

Fasting for a day on springwater is a great way to flush those toxins. Always check with your favorite healthcare provider before embarking on a fast. Also, if you get a headache, this is a natural reaction, so don't worry. If you really can't go all day without food, then eat fresh organic fruits and raw vegetables, plus lots of water.

December 27 - More Ways to Clean Your System and Feel Great!

Lemons help flush out your system, and help with nausea and heartburn. Sip a mug of hot water into which you've added the juice of half a lemon.

Bananas also soothe a stomach that has been pounded with rich food and alcohol.

December 28 - Resolve to Give Back

A great resolution for the New Year is to volunteer your time for others, even if just once a week, or once a month. Volunteering is all about commitment and enthusiasm, so choose something that you will enjoy and are interested in. Check at your local library, or read the local papers for organizations that need you. Here are some ideas to get you thinking:

- Teach an adult, or encourage a child, to read. Look for adult literacy classes at your library. Or ask at a senior residence if any residents would like you to read to them on a regular basis.

- Donate squares for afghans, shawls, lap quilts, and other handcrafted items for people suffering a loss or tragedy, who need to be comforted and reminded there are people who care.

- Meals on Wheels often needs drivers to deliver meals to the elderly and housebound.

- Your local animal shelter needs dog walkers, cat cuddlers, office staff, and more to help adopt out animal orphans.

- Your whole family can volunteer together planting trees, removing trash from beaches, rivers, and other beauty spots, or collecting data on dolphins, and so on.

- Be a mentor to a child.

December 29 - The Vegan Feast

If you're throwing a New Year's Eve party this year, whether for a few friends and neighbors or a whole house full, keep it a simple buffet, to keep you stress-free and sparkling. Here are some vegan ideas:

Has-Bean Paté

3	teaspoons olive oil
1	onion, finely chopped
1	teaspoon each of ground cumin and coriander
6	cups red beans, cooked
½	red bell pepper, chopped
2	cups fresh cilantro
	freshly ground black pepper, to taste

Be the Change You Want to See in the World

Heat the oil in a heavy skillet and gently sauté the onion and spices for a couple of minutes; do not burn. Stir in the beans and red pepper. Turn down the heat and cook for about 5 minutes, adding a little water if it gets too dry and sticks. Allow to cool a little, then blend in a food processor with the cilantro—keep a little lumpy. Check the seasoning and spoon into a serving bowl. Serve with slices of French baguette. Can be made the day before. Serves 4-6.

Fiesta Time!

1	can (15 ounces) vegetarian refried beans
3	teaspoons taco seasoning
3	tablespoons vegan sour cream
4	tablespoons vegan mayonnaise
2	avocados, mashed with a squeeze of lime juice
½	cup cilantro, chopped
2	green onions, finely chopped
1	cup vegan grated cheddar
2	large tomatoes, finely diced
1	cup sliced black olives, drained
	tortilla chips

In a pie plate or other suitable serving dish, spread the refried beans. Mix the taco seasoning with the sour cream and mayo, dollop over the beans, and spread out. On top of this, layer the avocado, cilantro, then the onions, half the cheese, the tomatoes, olives, and top with the rest of the cheese. Heat in a 350°F oven until the cheese just melts and the dish warms through a little. Serve with the chips. Serves 6-8.

December 30 - Gourmet Garbanzos

The Three Amigos Pasta Salad

1	can (15 ounces) kidney beans
1	can (15 ounces) garbanzo beans
1	can (15 ounces) black beans
½	package (14 ounces) of whole wheat pasta bows
1	bunch of green onions, chopped (or 1 red onion, thinly sliced)
1	sweet red pepper, seeded and finely sliced
1	head broccoli, cut into tiny florets
½	bunch radishes, washed and sliced
	salad dressing of your choice

Drain and rinse the beans and place in a large serving bowl. Boil the pasta. Cook to al dente, drain, and rinse with cold water. Add to the beans. Stir in the veggies and the dressing. Toss gently but well. Cover and place the bowl in the refrigerator for at least an hour before serving. Serves 6-8.

December 31 - Today is the First Day of the Rest of Your Life

And so another year ends. I hope this book has encouraged you to gather wonderful thoughts of your own and that they are recorded in your journal to help guide you through life.

Only you can dictate your happiness and your purpose in this world. By living consciously, your everyday decisions will make a difference to the outcome of your life.

There are many problems on this Earth, so finding ways to make significant, positive changes is certainly not difficult. Let's use our hearts to bring about brilliant transformations to everything that lives.

Remember, your world is what you make it, and it changes every day. Take control and make them happy changes by keeping your eyes, heart, and mind open. Free your spirit and use your energy wisely. Make your own dreams faithful to you.

I wish you a life of change making.

AFTERWORD

Do the world a favor—change your underwear.

Got your attention, didn't I? Several years ago, my company Tomorrow's World ran a few national ads with that headline to educate people about the benefits of organic cotton. Since advertising is intangible, it's hard to know exactly how productive it was, but I feel pretty confident it gave a chuckle to anyone who saw it. The hardest part of introducing new concepts is presenting the idea and providing the benefits. You have to inform readers how it will make their life better. The challenge in advertising our products is that not everyone realizes the benefits of using organic cotton (better for you and better for our planet). We need to rely on fact-driven knowledge instead of gimmicks to convey our message. It's harder and more costly to do this, but the long-term payoff for humankind is worth it.

In *Be the Change You Want to See in the World*, author Julie Fisher-McGarry presents a full year of educational tidbits for living a healthier lifestyle. Some of the ideas presented may be very easy to embrace, while others may be more challenging. As with most advice, you can take what you feel works for you and apply it to your life—simply modify those ideas that may not necessarily fit into your way of thinking. Fisher-McGarry writes with a friendly tone, in a diary style, making this book easy to read and comprehend. Have a pen and paper handy to take notes and try to incorporate her suggestions into your daily routine. There are many beneficial ideas throughout this book, and if you choose to embrace them you'll begin to notice little changes in your own environment. *Little changes that make a big impact.*

Each year more and more people understand the full-circle concept, how our decisions today affect our lives tomorrow. It could be something as drastic as a complete change of diet, or something as simple as changing your underwear. So, do the world a favor . . .

Cheryl Hahn, Founder and President of Tomorrow's World

www.tomorrowsworld.com

To Our Readers

Conari Press, an imprint of Red Wheel/Weiser, publishes books on topics ranging from spirituality, personal growth, and relationships to women's issues, parenting, and social issues. Our mission is to publish quality books that will make a difference in people's lives—how we feel about ourselves and how we relate to one another. We value integrity, compassion, and receptivity, both in the books we publish and in the way we do business.

Our readers are our most important resource, and we value your input, suggestions, and ideas about what you would like to see published. Please feel free to contact us, to request our latest book catalog, or to be added to our mailing list.

Conari Press
An imprint of Red Wheel/Weiser, LLC
500 Third Sreet, Suite 230
San Francisco, CA 94107
www.redwheelweiser.com